SMART ELECTRICAL AND MECHANICAL SYSTEMS

SMART ELECTRICAL AND MECHANICAL SYSTEMS

An Application of Artificial Intelligence and Machine Learning

Edited by

RAKESH SEHGAL
National Institute of Technology, Srinagar, India
NEERAJ GUPTA
National Institute of Technology, Srinagar, India
ANURADHA TOMAR
Netaji Subhas University of Technology, Delhi, India
MUKUND DUTT SHARMA
Department of Mechanical Engineering in lieu of Department of Electrical Engineering, National Institute of Technology, Srinagar, Jammu and Kashmir, India
VIGNA KUMARAN
Department of Electrical & Electronic Engineering, Universiti Tenaga Nasional, Malaysia

ELSEVIER

ACADEMIC PRESS
An imprint of Elsevier

Academic Press is an imprint of Elsevier
125 London Wall, London EC2Y 5AS, United Kingdom
525 B Street, Suite 1650, San Diego, CA 92101, United States
50 Hampshire Street, 5th Floor, Cambridge, MA 02139, United States
The Boulevard, Langford Lane, Kidlington, Oxford OX5 1GB, United Kingdom

ISBN: 978-0-323-90789-7

For information on all Academic Press publications visit our website at https://www.elsevier.com/books-and-journals

Acquisitions Editor: Lisa Reading
Editorial Project Manager: Sara Valentino
Production Project Manager: Prasanna Kalyanaraman
Cover Designer: Matthew Limbert

Typeset by TNQ Technologies

Contents

List of contributors

Vandan Agarwal
Microfluidics and Droplet Dynamics Lab, Department of Mechanical Engineering, Indian Institute of Technology, Indore, Madhya Pradesh, India

Parvaiz Ahmad
Department of Electrical Engineering, National Institute of Technology, Srinagar, Jammu and Kashmir, India

Mohammad Taghi Ameli
Faculty of Electrical Engineering, Shahid Beheshti University, Tehran, Iran

Ravi V. Angadi
Presidency University, Bengaluru, Karnataka, India

Vinay Arora
Computer Science & Engineering Department, Thapar Institute of Engineering and Technology, Patiala, Punjab, India

G. Arunkumar
Vellore Institute of Technology, Vellore, India

Goyal Awagan
Department of Electrical Engineering, National Institute of Technology, Raipur, Chhattisgarh, India

Sasan Azad
Faculty of Electrical Engineering, Shahid Beheshti University, Tehran, Iran

Kishore Bingi
School of Electrical Engineering, Vellore Institute of Technology, Vellore, Tamil Nadu, India

Puneet Chitkariya
Microfluidics and Droplet Dynamics Lab, Department of Mechanical Engineering, Indian Institute of Technology, Indore, Madhya Pradesh, India

Suresh Babu Daram
Department of EEE, Centre for Energy, Sree Vidyanikethan Engineering College, Tirupati, Andra Pradesh, India

C. Dhanamjayulu
Vellore Institute of Technology, Vellore, India

Anamika Dhillon
Department of Computer Engineering, National Institute of Technology, Kurukshetra, India

Subhojit Ghosh
Department of Electrical Engineering, National Institute of Technology, Raipur, Chhattisgarh, India

Neeraj Gupta
Department of Electrical Engineering, National Institute of Technology, Srinagar, Jammu and Kashmir, India

Sukanta Halder
Department of Electrical Engineering, Sardar Vallabhbhai National Institute of Technology, Surat, India

Rosdiazli Ibrahim
Department of Electrical and Electronics Engineering, Universiti Teknologi PETRONAS, Seri Iskandar, Malaysia

Nakka Jayaram
Department of Electrical Engineering, National Institute of Technology Andhra Pradesh, Tadepalligudem, India

Jerin John
Perigee Aerospace Inc, Daejeon, Republic of Korea

Pavan Kumar Kankar
System Dynamics Lab, Department of Mechanical Engineering, Indian Institute of Technology, Indore, Madhya Pradesh, India

P.S.V. Kishore
Department of Electrical Engineering, National Institute of Technology Andhra Pradesh, Tadepalligudem, India

Ebha Koley
Department of Electrical Engineering, National Institute of Technology, Raipur, Chhattisgarh, India

S.A. Lone
Department of Electrical Engineering, National Institute of Technology, Srinagar, Jammu and Kashmir, India

Ankur Miglani
Microfluidics and Droplet Dynamics Lab, Department of Mechanical Engineering, Indian Institute of Technology, Indore, Madhya Pradesh, India

Purushothaman Nandagopalan
School of Aeronautical Sciences, Hindustan Institute of Technology and Science, Chennai, Tamil Nadu, India

Amitkumar Patil
Malaviya National Institute of Technology, Mechanical Engineering, Jaipur, Rajasthan, India

Sakshee Patil
Microfluidics and Droplet Dynamics Lab, Department of Mechanical Engineering, Indian Institute of Technology, Indore, Madhya Pradesh, India

Anuj Prakash
Tata Consultancy Service, Bangalore, Karnataka, India

B. Rajanarayan Prusty
School of Electrical Engineering, Vellore Institute of Technology, Vellore, Tamil Nadu, India

Jami Rajesh
Department of Electrical Engineering, National Institute of Technology Andhra Pradesh, Tadepalligudem, India

Debanik Roy
Bhabha Atomic Research Centre & Homi Bhabha National Institute, Department of Atomic Energy, Mumbai, Maharashtra, India

Rakesh Sehgal
National Institute of Technology, Srinagar, India

Seyed Mohammad Shobeiry
Faculty of Electrical Engineering, Shahid Beheshti University, Tehran, Iran

Arshdeep Singh
Wipro Limited, Gurugram, Haryana, India

Gunjan Soni
Malaviya National Institute of Technology, Mechanical Engineering, Jaipur, Rajasthan, India

Shankarshan Prasad Tiwari
Department of Electrical Engineering, National Institute of Technology, Raipur, Chhattisgarh, India

Anuradha Tomar
Netaji Subhas University of Technology, Delhi, India

P.S. Venkataramu
Presidency University, Bengaluru, Karnataka, India

Gyanendra K. Verma
Department of Computer Engineering, National Institute of Technology, Kurukshetra, India

Eddie Yin-Kwee Ng
School of Mechanical & Aerospace Engineering, Nanyang Technological University, Singapore

CHAPTER ONE

Machine learning and its applications

Vinay Arora[1], Eddie Yin-Kwee Ng[2], Arshdeep Singh[3]
[1]Computer Science & Engineering Department, Thapar Institute of Engineering and Technology, Patiala, Punjab, India
[2]School of Mechanical & Aerospace Engineering, Nanyang Technological University, Singapore
[3]Wipro Limited, Gurugram, Haryana, India

Contents

Smart Electrical and Mechanical Systems
ISBN: 978-0-323-90789-7
https://doi.org/10.1016/B978-0-323-90789-7.00002-6

Machine learning (ML), an application of artificial intelligence (AI), is used for the purpose of data analysis which enables the companies to have a view of the existing trends in customer behavior and business operational patterns through the study of computer algorithms. It helps the companies to derive a competitive edge over their adversaries by developing new products. The automatic development of analytical model building is also based on this application. To understand the topic more clearly, it has been studied in detail under the following headings:

1. Artificial intelligence

AI is a replica of human intellect in computers, designed to believe and act like people. This might also mean any computer showing human-like characteristics such as learning and finding solutions. Machine learning (ML) is a subset of AI, where data are provided to computer and statistical strategies are used to help it "learn" how to succeed at a task without being specifically programmed for it. In ML, both unsupervised learning (containing unlabeled data sets) and supervised learning (with labeled data sets) are used. Deep learning (DL) is a kind of ML which analyses the data through a biologically inspired neural network design. The processing is done in neural networks through a number of hidden layers, enabling the machine to go "deep" with its learning, weighing input and forming connections for the best outcomes.[1] Fig. 1.1 depicts the relationship between AI, ML, and DL.

Artificial superintelligence (ASI), artificial general intelligence (AGI), strong intelligence, and narrow or weak AI are the types of AI. Weak or narrow AI represents a system capable of doing a single task. Weak AI systems comprise video games like chess and personal assistants like Amazon's Alexa. Strong AI, also known as AGI systems, performs human-like activities. These are often more sophisticated. They are trained to solve difficult problems without intervention of a human. These systems can be observed in applications such as hospital operating rooms or selfdriving automobiles. ASI is a hypothesized system that mimics or understands behavior and human intelligence in order to match human intellect and ability. In addition to replicating human intellect, ASI would potentially be far superior at everything a human person does, such as arithmetic, science, hobbies, athletics, medicine, art, emotional connections, and so on. ASI would have a larger memory and the capacity to process and analyze facts and stimuli more quickly. As a result, superintelligent beings would have considerably greater problem-solving and decision-making abilities than humans.

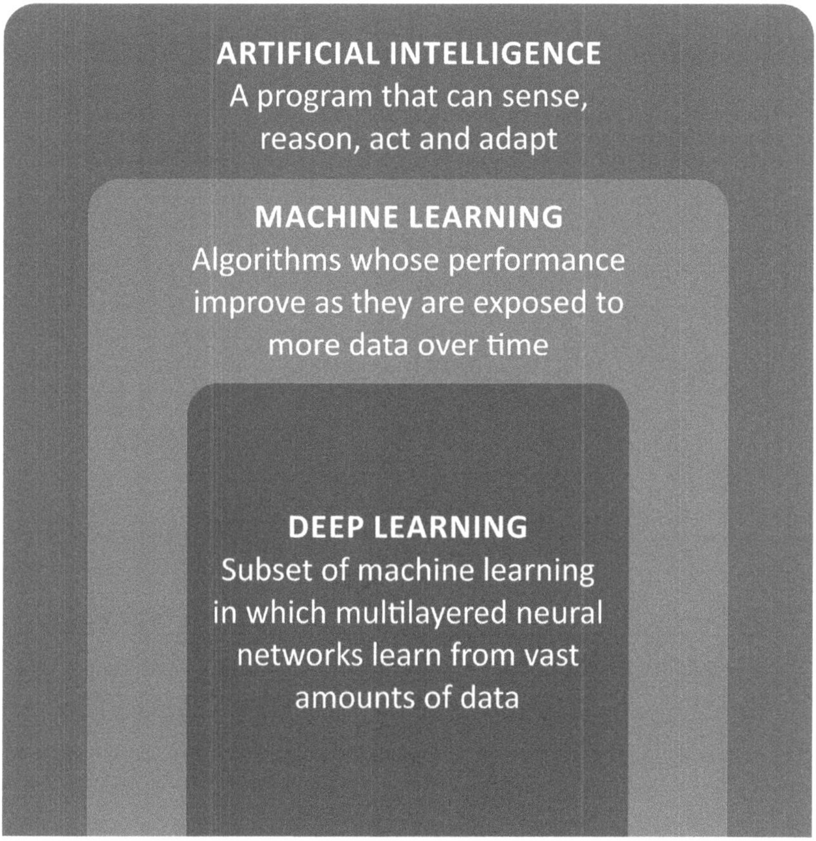

Figure 1.1 Relationship between AI, ML, and DL.[2]

Fig. 1.2 shows Sophia that is a science fictional entity made using the platform for improved robotics and AI research. Sophia is the world's first robotic citizen and the United Nations Development Program's first robot Innovation Ambassador. Sophia has made appearances on "Good Morning Britain" and "The Tonight" like shows, as well as at hundreds of conferences. Sophia is also a platform for AI and cutting-edge robotics research, with a focus on human–robot interactions, and the value for service and entertainment. Sophia has been utilized in studies as part of AI project, which aims to figure out how robots might adapt to their users' requirements via intra- and interpersonal growth.[4]

Figure 1.2 AI-based sophia robot.[3]

2. Machine learning and deep learning

ML is a branch of AI which allows computers to learn without being trained directly. A typical ML model initially learns from the data it encounters, then utilizes that knowledge to make predictions about new (future) data. The program is "trained" on a predefined collection of "training examples" or "training sets" in supervised ML. When a software is given data and asked to find patterns and correlations in it, this is known as unsupervised ML. A key advantage of ML techniques is the capacity to search for and discover patterns in vast amounts of data and in certain applications without prior knowledge. Fig. 1.3 illustrates the various components of ML.

DL is a feature of AI, which mimics the processing and creation of data in human brains to make decisions. DL is a branch of ML that studies unsupervised data from unstructured or unlabeled sources using neural networks. Deep neural learning, also known as deep neural network, is a form of ML that uses a hierarchical layer of artificial neural networks (ANNs) to complete tasks. ANNs are constructed in the same manner as the human brain is, with neuron nodes linked in a web-like framework. Unlike traditional programs, which produce analyses using data in a linear way, DL systems' hierarchical function allows computers to analyze data in a nonlinear fashion. Fig. 1.4A and B show internal scenario for the ML and DL models, respectively.

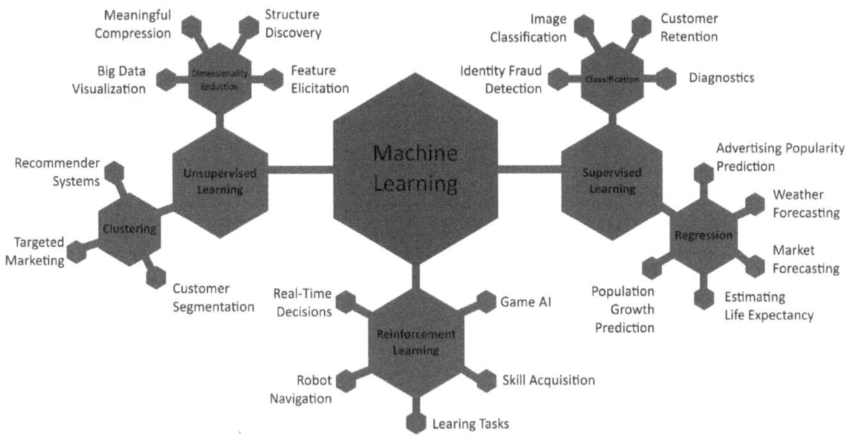

Figure 1.3 Machine learning at a glance.[5]

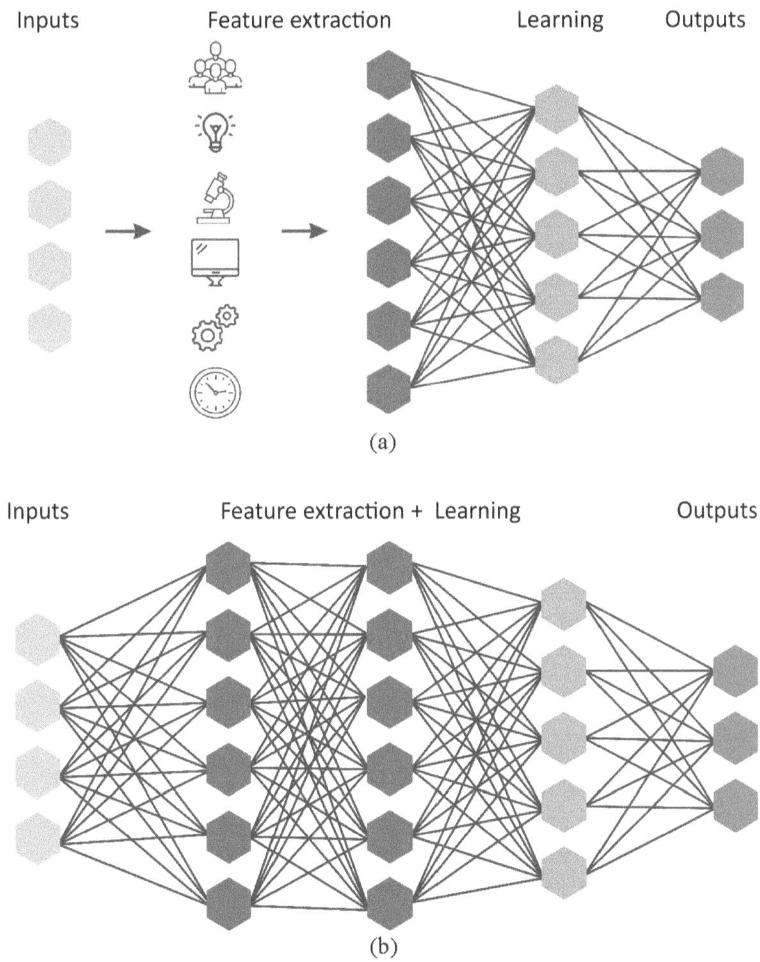

Figure 1.4 (A) Depiction of scenario for the machine learning and (B) deep learning.[6]

In traditional ML methods, the majority of appropriate features must be recognized by a domain expert in order to curtail data complexity and to make patterns more visible for the learning algorithms to work. DL algorithms aim to learn high-level properties from data gradually. This removes the requirement for domain knowledge and hard-core feature extraction.

Another significant distinction between ML and DL techniques is the approach to problem resolution. DL approaches address problems from beginning to finish, whereas the approaches of ML divide the problem statements into several portions first and then their findings are finally merged. For example, DL methods such as YOLO (You Only Look Once) net use an image as input and offer the position and name of items as output for a multiple object detection issue. However, in traditional ML methods such as support vector machine (SVM), a bounding box object identification technique firstly identifies all potential objects before using the histogram of oriented gradients as data to the learning algorithm to categorize relevant items.[4]

DL techniques are preferred due to the following reasons:

i. When the data amount is huge, the performance of DL is better than all other conventional approaches. On the other hand, traditional ML methods are preferred when there is a limited amount of data. DL methods need high-end infrastructure to operate in a reasonable amount of time.

ii. The performance of DL methods remains better than other methods when there is a lack of experience and expertise for feature introspection since feature engineering is less of a concern. DL handles with perfection even the complicated tasks like natural language processing (NLP), picture classification, and speech recognition.

3. Real life examples of deep learning and machine learning

This section provides some important real-world examples of ML and DL. Firstly, the application called Google Maps is among the today's most popular products, and its capacity to forecast traffic bottlenecks makes it vital for many drivers. According to Google, over one billion kilometers of road are traveled each day with the help of this application. To anticipate traffic patterns, Google and DeepMind researchers use data from multiple sources and integrate it into ML models. This data comprises live traffic information gathered anonymously from Android smartphones, local government information such as speed restrictions and construction sites, historical traffic data and variables such as the size, quality, and direction of every specific road. Secondly, Google Translate was made available as a statistical machine translation service. It translated the required material into English before translating into the target language. As of February 2010, it was incorporated into browsers like Chrome and could recognize words, pronounce text in images, and detect foreign text and languages.[4]

Thirdly, the most popular field has been personalization, in which ML empowers recommendation systems. Netflix has been using ML to help develop its collection of movies and television episodes by understanding what makes content effective. It is used to optimize the creation of original films and television series in Netflix's constantly expanding studio. ML also helps Netflix to improve adaptive bitrate selection, video and audio encoding, and its in-house Content Delivery Network. Fourthly, Tesla navigates highways and streets using cameras driven by computer vision algorithms. Deep neural networks are used by Tesla to recognize highways, objects, automobiles, and people via video streams from eight cameras mounted around the vehicle. DL contains distinct constraints that prevent it from comprehending the world in the same manner that humans do. Neural networks require massive quantities of training data to function consistently, and they lack the flexibility when presented with a novel circumstance that is not included in their training data.

Fifthly, image recognition is a widely used real-world application of ML. It can recognize an item as a digital picture based on the image pixels in black and white or color photos. The algorithm can find similarities and match them to faces by using a database of individuals. Sixthly, ML can convert voice to text. Live voice and recorded speech may both be converted into text files by some software tools. Intensities on time-frequency bands can also be used to partition speech. Devices such as Google Home and Amazon Alexa are the appropriate examples of how voice recognition software is used. Fig. 1.5A–F shows the various subdomains where ML/DL is providing assistance for developing the related applications/software.

4. Components of machine learning

4.1 Supervised learning

Supervised learning is a form of ML in which systems are trained with well-labeled dataset and then anticipate output based on those facts. The training data supplied to the computers in supervised learning, acts as a

Figure 1.5 Various arenas like (A) Google maps, (B) Google translate, (C) Recommendation systems, (D) Computer vision, (E) Image recognition, (F) Voice to text using cloud services, where machine/deep learning techniques are assisting to develop solutions.

supervisor, teaching the machines how to properly anticipate the output. It employs the same notion that a student would learn under the supervision of a teacher. Supervised learning is the process of supplying accurate input and output data to an ML model. A supervised learning algorithm aims to discover a mapping function that maps the input variable (x) to an output variable (y). Supervised learning, in real world, may be used for risk assessment, fraud detection, image categorization, spam filtering, and so on. Fig. 1.6 illustrates the concept used under the supervised learning.

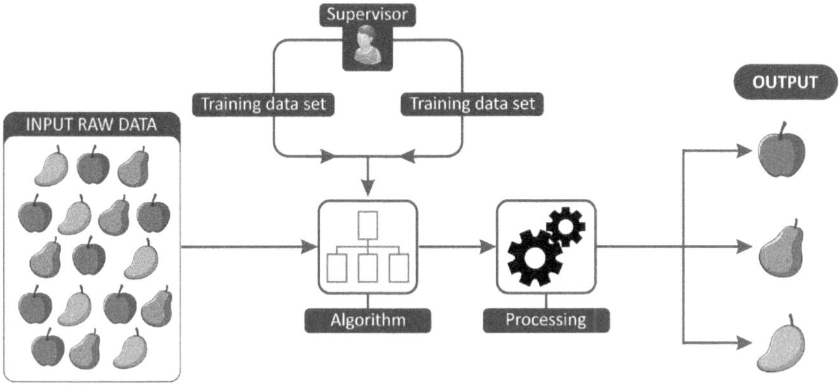

Figure 1.6 Mechanism used in the supervised learning.

Classification is a type of supervised learning where a computer software learns from data and makes new findings. Classification is the process of categorizing a given set of data into groups; it may be used taking both unstructured and structured data. Forecasting the class of supplied data points is the first step in this process. The classes are also called as the goal, categories, or label. Classification predictive modeling is used to calculate the mapping function between discrete input and output variables.[1] The principal goal is to determine which class/category the data will fit into.

The four different kinds of categorization tasks that one may experience are explained as below:

- *Binary Classification*: The term "Binary Classification" refers to classification jobs that contain two labels, for example, churn prediction, email spam detector, conversion prediction, *etc.* Typically, binary classification problems involve one class in its normal condition and another in its aberrant state.
- *Multiclass Classification*: This category includes categories with more than two class labels, for example, plant species classification, face classification, optical character classification, *etc.* In contrast to binary classification, multiclass classification does not incorporate the idea of normal and abnormal outcomes. Instead, instances are labeled as belonging to one of many recognized situations.
- *Multilabel Classification*: This is applicable to classification issues with two or more class labels, including one or more class labels expected for each instance. Let us take the example of image classification, in which a given photograph may have several things in the scene and a model may

predict the presence of various known objects in the image, such as "apple" "bicycle" "person" and so on.

- *Imbalanced Classification*: This relates to classification problems in which the number of instances in each class is not spread evenly. An unbalanced classification problem is often a binary classification job in which the majority of instances in the training dataset belong to the normal class or to the abnormal class, for example, fraud detection, medical diagnostic tests, outlier detection, *etc.*

Regression is a supervised learning approach which helps in the discovery of variable correlations and permits us to predict a continuous output variable using one or more predictor variables. It is mostly used for time series modeling, prediction, forecasting, and identifying the causal-effect connection between variables. In regression, we construct a graph between parameters that best matches the supplied data points, and the ML model may use this plot to draw conclusions about the data. Regression is simply a line or curve that goes through all of the data points on the target-predictor graph, reducing the vertical distance between them and the regression line. The gap between data points and lines indicates whether or not a model has correctly captured a strong link.[7]

The different forms of regression are explained as below:

- *Linear Regression*: The regression line has a linear shape; dependent variable is continuous; and the independent variable(s) can be either continuous or discrete. Linear regression employs a best-fit straight line to connect a dependent variable (Y) with one or more independent variables (X) (also called regression line).
- *Logistic Regression*: This should be used when the dependent variable is binary.
- *Polynomial Regression*: This method represents the connection of a dependent (Y) with an independent (X) variable as a polynomial of the nth degree. In ML, it is also known as a special instance of multiple linear regression since a few polynomial variables are inserted to the multiple linear regression equation to transform it to polynomial regression.
- *Ridge Regression*: It is implemented when the data has multicollinearity (independent variables having a high degree of correlation).
- *Lasso Regression*: Least Absolute Shrinkage and Selection Operator, like Ridge Regression, penalizes the absolute magnitude of the regression coefficients. Further, it has the potential to reduce variability and improve the accuracy of linear regression models.

4.2 Unsupervised learning

Unsupervised learning, an ML technique, creates models without the need of a training dataset. Models discover hidden patterns and insights in the provided data. It is similar to the learning that takes place in the human brain during learning new things. Unsupervised learning cannot be applied directly to a classification or regression issue. The uncontrolled learning aims to identify the fundamental structure of a dataset, which is grouped together in a compressed manner by similarity. Let us assume that a data collection with pictures of different sorts of cats and dogs is supplied for the unsupervised learning technique. The algorithm is not trained on the dataset; thus, it has no understanding of the characteristics of the dataset. The task of unsupervised learning algorithm is to detect the image characteristics by itself. This is done by grouping the picture dataset in groups according to similarities between the images.[1] Fig. 1.7 gives the working of unsupervised ML model.[7]

Clustering is a method of grouping objects so that those with the most similarities stay in one category, while those with few or no similarities stay in another. Cluster analysis finds similarities between data items and categorizes them based on the existence or absence of such similarities. An association rule is an unsupervised learning approach that is used to find associations between variables in a large database. It detects the group of objects in the collection that occur together. In Market Basket Analysis, the Association rule improves marketing strategy, for example, by indicating that people who buy X (say, bread) are more likely to purchase Y (butter/jam).

Dimensionality reduction refers to methods for decreasing the number of input variables in a dataset. It refers to the fact that adding more input characteristics to a predictive modeling assignment makes it more difficult to

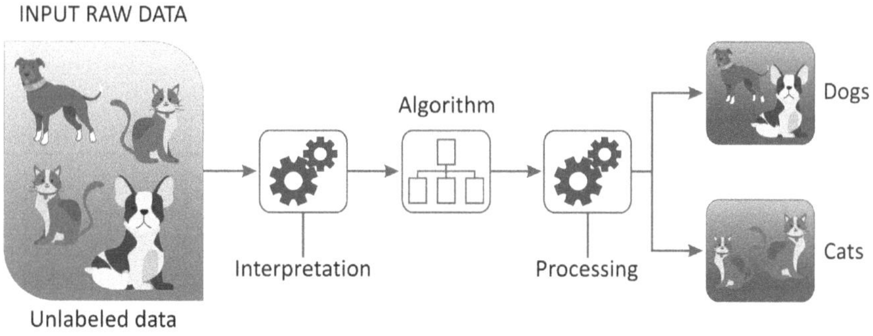

Figure 1.7 Working of unsupervised machine learning.[8]

model. Data visualization frequently makes use of high-dimensionality statistics and dimensionality reduction methods. These methods can be used in applied ML to reduce the size of a classification or regression dataset so that a prediction model can fit it better.

Advantage of Unsupervised Learning

- Unsupervised learning is preferred because unlabeled data are easier to get than labeled data.

Disadvantage of Unsupervised Learning

- The unsupervised learning method may be less accurate since the input data are not labeled and the program does not know the exact output in advance.

4.3 Reinforcement learning

As shown in Fig. 1.8, the reinforcement learning (RL) is a feedback-based learning method for the machine where the agent learns to comply through actions and outcomes in an environment. The agent receives positive feedback for every good move and negative feedback or punishment for every false move. In RL, unlike supervised learning, the agent automatically learns feedback without labeling data. The agent is compelled to learn from his experience in the absence of labeled information.[4]

RL handles a certain difficulty in sequential decision-making and aims for the long term, for example in gambling, robotics, and so on. The agent interacts and explores the surroundings on its own. The basic objective of reinforcement agent is to enhance performance by receiving the maximum positive rewards. The agent learns through the method of trial and error, and

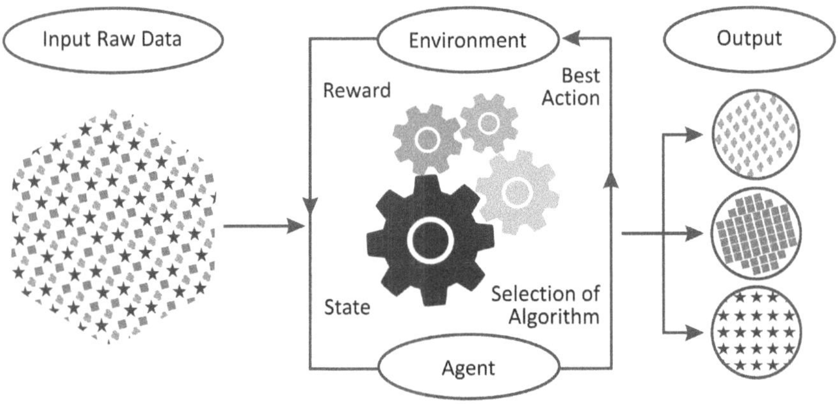

Figure 1.8 Working of reinforcement learning.[9]

as a result of its experience, it improves its ability to perform the task. Thus, it can be stated that "reinforcement learning is a sort of ML approach in which an intelligent agent (computer program) interacts with the environment and learns to operate within it."[7]

The four key components of RL are listed below:

(1) *Policy:* A policy describes the way an agent is required to act at a specific point of time. It connects environmental perceptions to actions performed in response to those perceptions. A policy is the only significant part of RL which decides the agent's behavior.

(2) *Reward Signal:* The reward signal determines the objective of RL. The environment delivers an instantaneous signal to the learning agent at each state, which is referred to as a reward signal. These bonuses are paid out based on the agent's positive and negative behavior. The agent's primary goal is to maximize the overall number of rewards for positive actions. The reward signal can modify the policy; for example, if an activity chosen by the agent results in a poor reward, the policy may change in the future to select different actions.

(3) *Value Function:* The value function indicates how beneficial the circumstance and action are, as well as the amount of reward an agent may expect. A reward denotes the signal for each good and bad action, but a value function denotes the future good condition and action. The value function is dependent on the reward since there might be no value if there is no reward. The estimating values aim to reap more rewards.

(4) *Model:* The model, which replicates the behavior of the environment, is the last component in RL. The model may be used to create predictions about how the environment will behave. For example, if a state and an action are provided, a model can predict the future state and reward.[4]

4.4 Ensemble methods

This sort of method is an ML methodology that combines many base models to generate a single optimum prediction solution. There are two types of ensemble techniques:

- *Sequential Ensemble Methods*: In this case, basic learners are created sequentially (e.g., AdaBoost). The primary purpose of using sequential techniques is to capitalize on the interdependence of the base learners. The overall performance can be improved by giving previously mislabeled instances more weight.

- *Parallel Ensemble Methods*: Here, base learners are created in parallel. Primarily, such techniques take advantage of the independence among the base learners since averaging may significantly minimize error.

Most of the ensemble techniques employ a single base learning algorithm to generate homogeneous base learners resulting in homogenous ensembles. Some techniques also employ heterogeneous learners, that is, learners of various kinds, resulting in heterogeneous ensembles. Ensemble methods demand the base learners to be as precise and diverse as feasible in order to be more accurate than their individual members. There exist three major categories of the meta-algorithms that aims at combining various ML techniques (viz. stacking, boosting, and bagging) into one predictive model:

4.4.1 Stacking

Fig. 1.9 presents an ensemble learning method that integrates many classifications or regression models using a meta-regressor or meta-classifier. It shows how to use a meta-learning approach to combine the predictions of two or more fundamental ML algorithms. On a classification or regression problem, stacking has the advantage of combining the abilities of several high-performing models to generate predictions that outperform any one model in the ensemble. Unlike bagging, the models in stacking are distinct (e.g., not all decision trees) and fit on the same dataset (e.g., instead of samples of the training dataset). Stacking, unlike boosting, utilizes a single model to learn how to most effectively integrate the findings from the contributing models (rather than a series of models that correct prior models' predictions).[1]

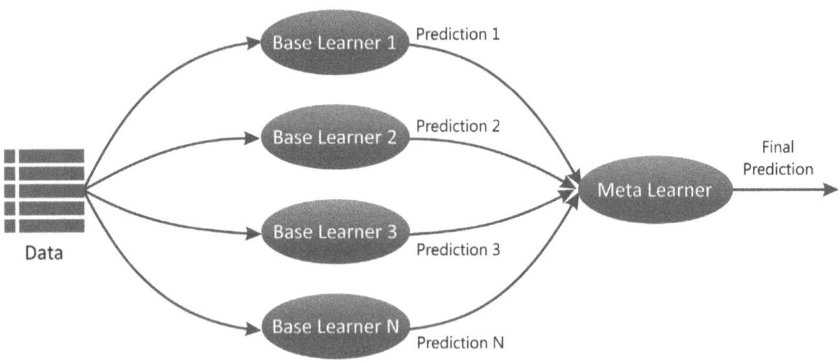

Figure 1.9 Steps to be followed for applying the stacking operation.[10]

4.4.2 Bagging

In Fig. 1.10, the Bagging classifiers are ensemble meta-estimators that fit base classifiers to random subsets of the given data and then combine their individual predictions (either by average or voting) to create a final prediction.

4.4.3 Boosting

This approach of ensemble modeling aims to create a strong classifier from a large number of weak ones. It is achieved by serially utilizing weak models to construct a model. Firstly, a model is created using the training data. Then, the second model is constructed with the aim to correct the flaws in the previous model. This method is repeated until either the whole training dataset is accurately predicted or the utmost of models are included. Each succeeding model seeks to fix the faults of the preceding model since boosting is a sequential process. Unlike bagging, it focuses on eliminating prejudice. Overfitting is a risk for boosting algorithms as a result of this. Some examples of boosting are eXtreme Gradient Boosting (XGBoost), Adaptive Boosting (ADABOOST), Gradient Boosting Machine (GBM), *etc.* Fig. 1.11 lists the steps involved under boosting process.[1]

4.5 Neural networks and deep learning

Neural networks are algorithms that recognize patterns and are loosely modeled after the human brain. They interpret sensory inputs by labeling or categorizing raw data using a kind of machine perception. All

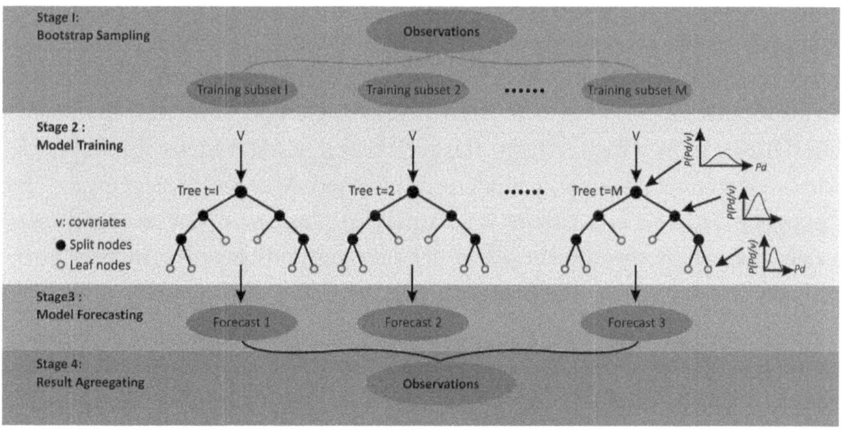

Figure 1.10 Internals of the bagging operation.[11]

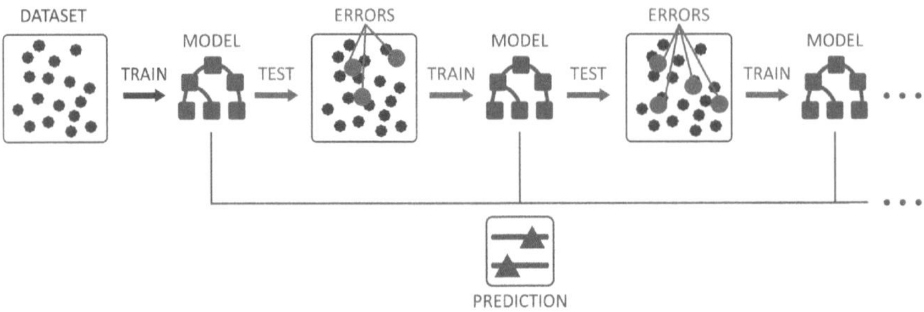

Figure 1.11 Steps under the boosting operation.[12]

real-world data, whether photos, text, music, or time series, must be converted into numerical patterns that may be stored in vectors. Neural networks can be used to categorize and cluster data. They may be seen of as a layer of grouping and classification on top of the data that can be stored and managed. They help in the grouping of unlabeled data based on similarities between example inputs, and classify data when given a labeled dataset to train on. Neural networks may also extract characteristics that are input into clustering and classification algorithms; so, deep neural networks may thus be thought of as parts of broader ML applications that include algorithms for RL, regression and classification. A Convolutional Neural Network (ConvNet/CNN) is a DL approach that can take an image as input, assign importance (learnable weights and biases) to various objects/aspects in the picture, and discriminate between them. When compared to other classification algorithms, a ConvNet requires much less preprocessing. While rudimentary techniques of filtering are used, ConvNets can learn these filters/characteristics with adequate training.[7]

Recurrent neural networks (RNNs) are the state-of-the-art technique for sequential data. These are used in Google's voice search and Apple's Siri. Due to its internal memory, it is well suited to ML issues with sequential data. Generative modeling is an unsupervised ML job that requires the discovery and learning of the regularities or patterns of incoming data in an automated method to create or produce new instances from which the original dataset might have plausibly been taken. Generative adversarial networks are a smart approach to learn a generative model by framing the problem with two submodels as a supervised learning problem: the model generator that we train to create new examples, and the model discriminator that attempts to categorize as genuine or false examples.

The encoder-decoder architecture can process inputs and outputs, both sequences with variable length, such that sequence translation issues, like machine translation, can be handled. The encoder is used as the input to turn the variable-length sequence into a fixed form state. The second component is a decoder and it translates a changeable sequence of the encoded condition of a fixed form. A perceptron is an artificial neuron (neural network unit) which detects functionality or business intelligence in the inserted data using particular algorithms. In 1957, Frank Rosenblatt launched perceptron. Frank presented a rule based on the original McCulloch-Pitts Neuron for perceptron learning. Perceptron is a supervised binary classification learning method, which allows neurons to learn and process information one at a time during training. Perceptron is of two types: single layer and multilayer. Perceptron can only learn patterns that are linearly separable. The higher processing capacity is achieved by the multilayer perceptron (MLP) or neural feedforward network having two or more levels. The perceptron learns the mass of the incoming signals so that a linear boundary can be established.

A feedback neural network is a biologically inspired categorization method. It consists of a number of fundamental neuron-like processing units which are laid together (possibly enormous). Each unit of a layer is connected to all units of the previous layer. All of these connections are not identical; each might differ in weight or strength. The strength of these links encodes network information. The nodes are often known as the units in the neural network. The data enter as input and travel across the network layer after layer until the outputs are reached. During normal functioning there is no feedback between layers, which is when it works as a classifier. This is why they are known *feedforward* neural networks (FFNN). Long Short-Term Memory (LSTM) networks are a type of RNN which may learn order dependency when performing sequence prediction tasks. In many aspects, traditional feed-forward neural networks and RNNs are outperformed by LSTMs. This is because of the limited capacity of FFNN and RNN to selectively recall patterns across long periods of time.[7]

5. Application areas of machine learning and deep learning

5.1 Smart manufacturing

Smart manufacturing seeks to integrate big data, Industrial Internet of Things, high-performance computing, and sophisticated analytics into traditional manufacturing structures and procedures in order to create highly customizable, high-quality products at lower costs. Unlike a traditional factory, a smart factory uses intelligent automation systems, interoperable information and communications technologies (ICT), and sensor networks to monitor machinery conditions, diagnose root causes of failures, and anticipate the remaining useful life (RUL) of mechanical components or systems. Almost all engineering systems (e.g., machine tools, nuclear power plants, and aircraft systems) are susceptible to mechanical breakdowns due to wear and tear, age, or abnormal operating circumstances. Excessive load, overheating, fracture, deflection, corrosion, fatigue, and wear are all common failure modes. Due to unanticipated machine downtime, the deterioration and failure of engineering systems or components will often result in greater expenses and decreased output. Thus, it is important to create and implement an intelligent maintenance plan that allows manufacturers to assess the health of in-service equipment and anticipate when maintenance should be conducted in order to enhance production efficiency while lowering maintenance costs.[13] Tool wear refers to the progressive breakdown of cutting tools as a result of frequent use. Tool bits, tipping tools, and drill bits for machine tools are among the tools that are impacted. Predictive modeling may be used to monitor and predict tool wear situations. Three prominent ML methods, viz. ANNs, support vector regression (SVRs), and random forest (RFs) are used to forecast tool wear in milling operations.

The RUL is an estimate of how many more years an item, component, or system is expected to perform in line with its original purpose before it has to be replaced. Vanilla LSTM neural networks are used to tackle the bottleneck problem of RUL estimation for sophisticated engineering systems, which generally perform well in the field of NLP. The RUL of equipment that is subjected to condition monitoring has also been predicted using ANN.

One of the most frequently used components in rotating equipment is the rolling bearing. It is so easy to damage that it might result in a mechanical failure. A rolling bearing is an essential component of rotating equipment, and the status of its operation has a significant impact on the equipment's overall health. To avoid disastrous machine malfunctions, loss of output, and human casualties, it is critical to monitor the state of rolling bearings and diagnose their defects. Rolling bearing faults may be diagnosed using Stacked Auto-Encoders and Deep Belief Networks[14] and motor bearing using neural networks.[15] Table 1.1 summarizes the research contributions related to the role of ML/DL in the subdomain of smart manufacturing.

Table 1.1 Research work related to smart manufacturing and machine learning.

	Related task	ML/DL model	Citations
Smart manufacturing	Prediction of tool wear	SVR	13,16,17
		ANN	13,16,18–20
	Remaining useful life prediction	LSTM NN	21
		ANN	22,23
	To diagnose the fault of rolling bearing	Deep belief network	14,24
		Neural networks	15

5.2 Robotics and autonomous vehicles

Robotics is an interdisciplinary field whose aim is to create devices that help and support people. Mechanical engineering, electrical engineering, information engineering, mechatronics, computer engineering, bioengineering, electronics, software engineering, and control engineering are some of the areas that robotics encompasses. Robotics is based on developing machines that can replace humans and replicate their actions. Robots may be employed in a variety of circumstances and for a variety of reasons; however, many are now deployed in hazardous areas (including bomb detection, radioactive materials, and deactivation), manufacturing processes, or situations in which humans find it hard to live.

Autonomous automobiles, unlike autopilot, do not require human contact or control. These systems are intended to exclude humans from the equation entirely. These systems use radar, sensors, cameras, and computer vision control systems to perceive their surroundings and interpret everything from road conditions to traffic signs. Autonomous systems use a number of different algorithms to analyze data and assess risk. They may, for example, identify a youngster on a sidewalk, establish that it is a child, and analyze whether the child's actions will put it in the vehicle's path. If the computer detects the possibility of a collision, it will either slow, halt, or alter the vehicle's motions to prevent colliding.[25] DL has gained popularity as a consequence of several state-of-the-art achievements in areas like image classification and speech recognition. Consequently, DL is being used most commonly in autonomous vehicle applications such as decision-making and planning, vision, and mapping and localization.[26]

CNN performance with raw video inputs has the ability to limit the number of sensors required by selfdriving cars. As a result, several automobile companies are looking at autonomous cars which do not rely on pricey sensors like Light Detection and Ranging and instead rely on DL for object recognition, scene analysis, motion prediction and semantic segmentation. The outstanding performance of DL in these perceptual problems has sparked interest in utilizing DNNs to create control actions in selfdriving automobiles.[27] DL helps greatly in vehicle control. It is best suited to control difficulties in complex and dynamic contexts due to its capacity to selfoptimize its action from data and adapt to new situations. Instead of adjusting each parameter sequentially while striving to preserve performance in all conceivable circumstances, DL allows developers to define the intended behavior and train the system to perform nicely as per the new conditions. For these reasons, DL has attained greater attention in the recent years.[25]

There are five levels or degrees of driving automation, each indicating the vehicle's ability to act and respond independently. These are described as below:

- Level 0: There is no automation.
- Level 1: Driver Assistance—In some instances, the car can handle speed or steering independently to help the driver.
- Level 2: Partial Automation—In some conditions, the car may manage both speed and steering independently to help the driver.
- Level 3: Conditional Automation—Under typical environmental circumstances, the car can manage both steering and speed automatically, although it still requires human intervention.
- Level 4: High Automation—Under typical environmental circumstances, the car can complete a journey without the need for driver intervention.
- Level 5: Full Autonomy—In every setting, the car can finish a journey independently.

In terms of vehicle autonomy, Tesla is at Level 2. Deep neural networks are used to recognize highways, automobiles, objects, and people via video streams from the vehicle's eight cameras. PyTorch has been used to train networks in Tesla.[28] For object identification in the new Audi A8, supervised learning has been utilized. Deep neural networks are trained utilizing a variety of datasets throughout this process. The neural network learns to categorize a wide range of items in this way, such as automobiles, bicycles, and people. The data gathered during this method are then made available to the finished product of the driver assistance system software and also piloted driving software.[29]

The development of an effective human gesture recognition system is required for autonomous cars to avoid collisions. Infrared photography, as opposed to visible spectrum cameras, can provide more reliable human gesture detection in a complicated context. To improve human gesture detection and convolutional feature propagation in the YOLO-V3 (You Only Look Once-Version 3) network, three DenseNet blocks have been added before the residual components.[30] The most common method for incorporating extremely detailed, centimeter-level maps into behavioral prediction models is to convert the map into pixels and use a CNN to encode all of the scene information, such as crosswalks, road lanes, traffic signs, and road borders. However, this technique consumes a lot of CPU power and takes time (latency), which is inconvenient for a selfdriving car that needs to make judgments in milliseconds.[31]

The DL methods are disadvantageous due to huge quantity of data and time necessary for proper training, particularly for RL methods. This can result in long learning periods, causing delays and increased costs in the development of autonomous vehicles. Combining RL and supervised learning that helps to minimize training time while still offering excellent adaptability is a typical approach for reducing training data needs or training time. Due to the large amount of data that must be collected, training a vehicle to drive in all potential circumstances that it may face in the real world is difficult. The threat of overfitting to the dataset grows when the amount of data used to train the model is raised without guaranteeing diversity in the dataset.[25] Table 1.2 summarizes the research contributions related to the role of ML/DL in the subdomain of robotics and autonomous vehicles.

5.3 Smart electric power system

A smart energy system is a long-term, secure, and cost-effective system that coordinates and integrates renewable energy generation, infrastructure, and consumption via active users, energy services, and supporting technology.[36]

Table 1.2 Research work related to robotics and autonomous vehicles.

	Related task	ML/DL model	Citations
Robotics and autonomous vehicles	Autonomy in 9	Deep neural networks	28
	Autonomy in audi A8		29
	Autonomy in waymo	VectorNet	31
	Human gestures recognition for autonomous vehicles	DenseNet	30,32,33
	Super resolution (direction-of-arrival) DOA estimation	SBLNet	34,35

Current power systems are rapidly transitioning to their smart grid counterpart which is more flexible, active, and intelligent, posing enormous challenges in a variety of areas, including cyberspace security, system planning decision-making, demand-side management, and operation. Advanced smart grid functionality is dependent on the underlying communication infrastructure and information, as well as the effective handling of vast amounts of data provided by numerous sources, such as phasor measurement units, smart meters, and other types of sensors.[36]

The power systems' dependability, security, and efficiency may all benefit from this bidirectional power and data flow. As a result of its enhanced energy efficiency, regulated CO_2 emissions, cost-effectiveness, improved utility, and lower cost, the smart grid is considered as the future power grid.[36] Digital technology enables the smart grid, allowing for two-way contact between the utility and its consumers, as well as monitoring along transmission lines. The Smart Grid, like the Internet, will be composed of controls, automation, computers, new technologies, and equipment that operate in tandem; however, these technologies will interact with the electrical grid to digitally adapt for changing energy demands.[37]

The advantages of Smart Grid are as here under:

- Improved power transmission efficiency.
- Faster restoration of power in the event of a power outage.
- Reduced utility operation and management expenses, resulting in lower electricity bills for customers.
- Improved security.
- Lowering electricity rates due to decreased peak demand.
- Customer-owned power generation systems, particularly renewable energy systems, have been improved.

False data injection attack detection concerns in the smart grid are posed as statistical learning problems for numerous attack scenarios, with measurements taken in batch or online settings. k-Nearest Neighbor, Perceptron, SVM, and Sparse Logistic Regression are the 4 ML supervised methods that have been be used to identify attacks in the smart grid environment.[38] The smart grid is a system that makes the electrical grid more efficient, dependable, and secure. It has been possible due to recent advancements in sensing, information, and communication technology. An energy resource management plan must be established to balance energy demand and supply in the smart grid in order to provide energy resources efficiently. SVM and Neural Networks are being utilized to make accurate predictions for using energy.[39]

The communication infrastructure not only connects the important power facilities but also addresses certain cyber security issues. The communication channels that are utilized in state estimate may be vulnerable to cyber-attack. A cyber-attack in state estimators might be triggered by a variety of reasons, including (1) making financial benefits; (2) causing technical issues with the grid, such as blackouts; and (3) a mix of financial advantages and technological difficulties. Analyzing vulnerabilities in state estimates (and, more broadly, in the smart grid) allows control centers to enhance operational safety by evaluating appropriate countermeasures. The smart grid system is used to identify stealth assaults with SVM in[40]. Table 1.3 summarizes the research contributions related to the role of ML/DL in the subdomain of smart electric power system.

5.4 Image processing in mechanical engineering and electrical engineering

Image processing is the application of procedures to an image in order to enhance it or derive useful information from it. It is a type of signal processing where an image is supplied; and the output is either the picture or its characteristics/features. The two types of image processing technologies utilized are analog and digital image processing. Analog image processing can enhance hard copies such as prints and images. Image analysts use a range of interpretative foundations while using these visual methods. Digital image processing techniques enable computer-assisted picture alteration. The three major phases that all types of data must go through when employing digital technology are preprocessing, augmentation and display, and information extraction.[47]

Table 1.3 Research work related to the smart electric power system.

	Related task	ML/DL model	Citations
Smart electric power system	Attack detection in smart grid	Perceptron, k-nearest neighbor, sparse logistic regression	38
		SVM	38,40–42
	Detection of compromised smart grid devices	Random forest	43
	Forecasting of energy usage	SVM and NN	39,44
	Load classification in smart grid	K-means	45
		FCM	45,46

Mechanical engineering is a field of engineering basically concerned with the industrial application of mechanics and manufacturing of tools, machines, and their products. Power plant engineering, manufacturing engineering, thermal engineering, automobile engineering, and mechatronics engineering (a mix of mechanical, electrical, and computer engineering) are all examples of mechanical engineering disciplines.[48] Electrical engineering is the study of electrical control systems, machinery, and equipment, as well as its design, construction, and maintenance.[49] Detecting and measuring defects on metal and metal-coated surfaces is one of the most difficult elements of computer vision-based semantic segmentation. When dealing with a high number of test subjects, manual inspection is not a viable option for finding flaws such as surface-level fractures, scratches, and manufacturing errors. Metal flaws come in a variety of sizes, shapes, and textures, and they frequently resemble potential artifacts owing to regular wear and tear and metal coating brush marks. As a result, developing an effective segmentation approach for measuring such flaws will be difficult. Metal fractures have been identified and segmented using Deep CNN in[50] and to identify and diagnose material utilizing microscopic pictures in[51].

Pantograph is one of the main elements of electric rail cars. The sliding plate's surface must be as light as possible to provide a continuous train delivery of power from one side to the other with minimum friction. Furthermore, for safety reasons, the pantograph sliding plate thickness cannot be less than the lowest limit. Periodic examination and repair of the pantograph sliding plate is therefore critical for safe and stable functioning. The Probability Density Displacement Network (PDDNet) has certain benefits over AlexNet due to its use of series convolution layers.[52] In the surface faults detection of the pantograph slider, an upgraded Faster R—CNN algorithm was employed, which surpassed the limits of existing detection techniques.[53] Table 1.4 summarizes the research contributions related to the role of ML/DL in the subdomain of image processing in mechanical engineering and electrical engineering.

5.5 Renewable energy

Wind power, solar energy, hydropower, tides, waves, biomass energy, and geothermal energy are the examples of renewable energy. Renewable energy is a type of energy that is environmentally friendly. This carbon-free technology is helpful in the fight against climate change and has emerged as a viable alternative to today's petrochemical energy sources. Renewable

Table 1.4 Research work related to the image processing in mechanical engineering and electrical engineering.

	Domain	Related task	ML/DL model	Citations
Image processing	Mechanical engineering	Detection and classification of surface defects of gun barrels	SVM	54
		Classification of roller bearing faults		55,56
		Localization and segmentation of metal cracks	Deep CNN	50
		Detection and diagnosis of material degradation		51
		Defect detection	PDDNet	52
		of pantograph	Improved R−CNN	53
	Electrical engineering	Image reconstruction in electrical resistance tomography (ERT)	CNNs	57
		Electrical load forecasting	SVM	58

energy, on the other hand, has a wide range of properties, which leads to uncertainty in renewable energy power systems. As a result, forecasting renewable energy is an essential approach to address this issue.[59] To date, a variety of techniques have been developed to increase the prediction performance of renewable energy, including statistical methods, physical models, AI techniques, and their hybrids. DL is one of them since it is a potential way to find nonlinear and high-level invariant data structure in ML.[60]

Among renewable energy sources, wind energy is a significant source that has the ability to provide electricity in a sustainable and continuous manner.[61] In wind speed prediction, ML methods are employed. As winds show temporal changes in their application of a number of magnitude levels, like annual changes (due to climate changes), seasonal changes, daily variations (due to the local microclimate), short fluctuation (due to land and sea breezes), and hourly fluctuations (due to land and sea breezes) (bursts). When wind turbines are subjected to inclement weather, they suffer financial losses. As a result, even in the short future, it is important to develop accurate techniques for predicting wind speed.[62]

A photovoltaic (PV) module is a group of PV cells installed in a framework. PV cells provide direct power with sunlight as a source of energy. A PV Panel is the PV modular collection, whereas an Array is a panel collection. The arrays of a PV system supply electricity from solar energy. The use of CNN-based techniques has the ability to forecast PV system maintenance. Forecasting solar energy is another use of ML for which we utilize LR, ANNs, RF, and SVR.[63]

In recent years, the factors such as rising worldwide energy demand, fossil fuel depletion, and growing environmental concerns about fossil fuels have prompted academics to look for ways to replace fossil fuels with clean energy sources based on renewable resources. Biomass and hydrogen have fetched a lot of attention among renewable energy sources since they can help enhance global energy sustainability while also lowering greenhouse gas emissions. Biomass gasification is among the most promising methods for extracting energy from agricultural, forestry, and livestock farming by-products and wastes.[64] The MLP and Decision Tree Regression (DTR) algorithms have been used to forecast the outcome of gasification as CO, CO_2, H_2, CH_4, or HHV.[65]

The rise and fall of water in the ocean produce tidal energy, which is a sustainable source of electricity. Lower power losses, simple power supply for offshore residences and businesses, minimal emission, rapid replenishment, and easy accessibility are some of the advantages of employing tidal power in electrical systems. It is critical to accurately estimate and anticipate tidal power in order to obtain the most efficiency out of this renewable energy source. As a hybrid model, SVR has been utilized to reliably forecast tidal current speed and direction.[66] Tidal ranges have a major impact on ocean waves, offshore structures, sediment movement and coastal aquifers, in addition to their impacts on beach profiles. A novel deep neural network has been suggested to forecast tide levels for various places at any time.[67] Table 1.5 summarizes the research contributions related to the role of ML/DL in the subdomain of renewable energy.

The intensity of the wake turbulence is critical to assess since it has a direct influence on the fatigue loading on the wind turbine blades.[94] In their study, Kabir and Ng[95] had validated the Computational Fluid Dynamics (CFD) results of the National Renewable Energy Laboratory (NREL) Phase

Table 1.5 Research work related to the renewable energy.

	Domain	Related task	ML/DL model	Citations
Renewable energy	Wind	Predicting wind speed	ANN	68–73
			SVM	74–78
			SVR	79,80
	Solar	Predictive maintenance of PV system	LR, RF, SVR, ANN	63
		Estimation of solar radiation	SVM, SVR, k-means, RF, RT	81
			ANN	81–83
	Biomass	Gasification	MLP and DTR	65
		outcome prediction	NN	84
		Forecasting yields of crop	ANN	85–87
	Wave	Maximum energy absorption	NN	88
		Prediction on structural stress and deformation of fish cage in waves	ANN	89
		Calculations of peak of typhoon waves		90
	Tidal	Prediction of tidal current	SVR	66,91,92
		Accurate tide level estimation	DNN	67,93

VI wind turbine against the experimental measurements. To further generate high-fidelity training and test datasets for the ML models and to establish the underlying complex relationship between the input and the output variables, CFD simulations are carried out for the NREL Phase VI wind turbine[94,95] as well as the MEXICO wind turbine rotor Phase-III.[96]

Kabir et al.,[97] who developed wake models using Genetic Programming where turbulence intensity has been assessed utilizing ML modeling.[94,95] Three ML algorithms, namely, SVR, ANN, and Extreme Gradient Boosting (XGBoost), have been validated to estimate the velocity and turbulence intensity of a wind turbine's wake at distinct downstream distances.[98] The results demonstrate that ML-based algorithms can predict velocity and turbulence intensity better with higher precision than the traditional analytical wake models.[98]

Table 1.6 Research work related to the optimal power flow.

	Related task	ML/DL model	Citations
Optimal power flow	Solving direct current optimal power flow (DC-OPF) problems	DNN	102,103
	Predicting future of ACOPF problems	Random forest	100
	The economic load dispatching (ELD) problem	Genetic algorithm	104–106
	To find the optimal power that the generators within the grid have to produce to satisfy a given demand.	Graph neural networks	101
	To predict active and inactive constraints	Support vector classifiers	107

5.6 Optimal power flow

Many power system operating tools rely on optimal power flow (OPF).[99] The OPF problem, initially stated by Carpentier in 1962, aims to minimize all operational, physical, and technological limitations. The OPF issue is critical to power system management because it underlies a variety of applications such as unit commitment, economic dispatch, reliability, demand response, and stability evaluation.

Grid operators addressing AC optimum power flow (ACOPF) have created a huge quantity of data over the years. The data may be utilized to forecast future ACOPF problems using a RF model.[100] OPF tries to figure out how much electricity the generators in the grid should supply to meet a particular demand. The OPF solution was approximated using graph neural networks.[101] Table 1.6 summarizes the research contributions related to the role of ML/DL in the subdomain of OPF.

5.7 Industrial control system

Industrial control system (ICS) is a collection of control systems and associated tools that encompass the instruments, networks, systems, and controls utilized in the operation and/or automation of industrial processes. Each ICS is meant to operate electronically and efficiently depending on the

Table 1.7 Research work related to the industrial control system.

	Related task	ML/DL model	Citations
Industrial control system	Detection of cyber attacks	DNN	110,111
	Anomaly detection	DT	112,113
	Intrusion detection system	Multilayer perceptron	114,115
		(Robust incremental principal component Analysis) RIPCA-(one-class support vector machine) OCSVM	116
	Defense against process-aware attacks	SVM	109,117

company. In almost all industries and vital infrastructure such as manufacture, energy, transportation, and water treatment, ICS devices and protocols are now present. Numerous types of ICSs are available, including the most common monitoring supervisory control and data acquisition and distributed control systems.[108]

ICS is experiencing new security issues and is vulnerable to the same vulnerabilities that afflict IT, as seen by the growing number of ICS-related incidents. One can train an SVM model which can identify anomalies in real time and differentiate between disruptions and malicious behavior.[109] Table 1.7 summarizes the research contributions related to the role of ML/DL in the subdomain of ICS.

6. Benchmark dataset(s)

6.1 Kaggle

Kaggle is a website that enables data science and other designers to organize, produce, and exchange code and host data sets for ML contests. Data science tasks on Kaggle range from attempting to forecast cancer recurrence by reviewing health information to analyzing the mood evoked by movie reviews and how this influences audience reaction. This trailblazing platform accepts projects from a variety of sources. Some are purely instructive and enjoyable brain exercises, while others are actual problems that businesses are attempting to tackle.[118]

6.2 Dataset search

The goal of this mark-up is to make datasets from disciplines including biological sciences, ML, social sciences, civic and government data, which will also be easier to find. Google's Dataset Search tool has been used to search for the datasets.[119]

7. Conclusion

The necessity for advances in power system design, operation, and control is dictated by recent innovations in the power system domain. In today's ever-increasing competitive environment, the power equipment business faces new problems. AI methods, when combined with traditional analytical tools, can make a major contribution to the resolution of connected issues, such as the ongoing reform of energy markets, the rising number of interconnections and power exchanges among utilities, the growing need for power system efficiency and power quality, *etc.* The major user of AI as a technology is the Mechanical Engineering industry. It is used in mechanical designs or engineering operations in greater quantities than in any other business. AI has been used in sections of Mechanical Engineering such as Robotics, Automation, and Sensor Technology. As a result, it is simple to claim that Mechanical Engineering disseminates AI application and utilization in the ecosystem.

References

1. Wlodarczak P. *Machine Learning and its Applications.* Taylor & Francis Limited; 2021. ISBN 9781032086774, 1032086777.
2. Stackexchange, https://ai.stackexchange.com/questions/15859/is-machine-learning-required-for-deep-learning [as accessed on 2 July 2021].
3. Meriame Berboucha, https://www.forbes.com/sites/meriameberboucha/2018/11/29/little-sophia-a-new-robot-citizen-has-entered-our-world/?sh=e3d5d9a39007 [as accessed on 4 July 2021].
4. Bhattacharyya S, Das R, Nandy S. *Machine Learning Applications.* De Gruyter; 2020. ISBN 9783110608663, 3110608669.
5. D. Johnson, https://www.guru99.com/machine-learning-tutorial.html [as accessed on 2 September 2021].
6. Quantdare, https://quantdare.com/what-is-the-difference-between-deep-learning-and-machine-learning/deep_learning/ [as accessed on 4 June 2021].
7. Hiran KK, Meena G, Mahrishi M, Sharma P. *Machine Learning and Deep Learning in Real-Time Applications.* IGI Global; 2020. ISBN 9781799830979, 1799830977.
8. Javatpoint, https://www.javatpoint.com/unsupervised-machine-learning [as accessed on 10 July 2021].
9. Techvidhvan, https://techvidvan.com/tutorials/reinforcement-learning/ [as accessed on 5 July 2021].

10. Analyticsvidhya, https://www.analyticsvidhya.com/blog/2021/06/ensemble-deep-learning-an-ensemble-of-deep-learning-models/ [as accessed on 2 September 2021].
11. Diogo Menezes Borges, https://www.kdnuggets.com/2019/01/ensemble-learning-5-main-approaches.html [as accessed on 3 June 2021].
12. Bradley B., https://bradleyboehmke.github.io/HOML/gbm.html [15 June 2021].
13. Wu D, Jennings C, Terpenny J, Gao RX, Kumara S. A comparative study on machine learning algorithms for smart manufacturing: tool wear prediction using random forests. *J Manuf Sci Eng*. 2017;139:071018. https://doi.org/10.1115/1.4036350.
14. Chen Z, Deng S, Chen X, Li C, Sanchez R-V, Qin H. Deep neural networks-based rolling bearing fault diagnosis. *Microelectron Reliab*. 2017;75:327−333. https://doi.org/10.1016/j.microrel.2017.03.006.
15. Li B, Chow M-Y, Tipsuwan Y, Hung JC. Neural-network-based motor rolling bearing fault diagnosis. *IEEE Transact Indus Electron*. 2000;47:1060−1069. https://doi.org/10.1109/41.873214.
16. Wu D, Jennings C, Terpenny J, Gao R, Kumara S. Data-driven prognostics using random forests: prediction of tool wear. In: *Proceedings of ASME 12th International Manufacturing Science and Engineering Conference, MSEC Collocated with the JSME/ASME 6th International Conference on Materials and Processing, Los Angeles, United States*. June 8 , 2017. https://doi.org/10.1115/MSEC2017-2679. V003T004A048.
17. Wang J, Wang P, Gao RX. Enhanced particle filter for tool wear prediction. *J Manuf Syst*. 2015;36:35−45. https://doi.org/10.1016/j.jmsy.2015.03.005.
18. Palanisamy P, Rajendran I, Shanmugasundaram S. Prediction of tool wear using regression and ANN models in end-milling operation. *Int J Adv Manuf Technol*. 2008;37:29−41. https://doi.org/10.1007/s00170-007-0948-5.
19. Twardowski P, Wiciak-Pikuła M. Prediction of tool wear using artificial neural networks during turning of hardened steel. *Materials*. 2019;12:3091. https://doi.org/10.3390/ma12193091.
20. Sanjay C, Neema M, Chin C. Modeling of tool wear in drilling by statistical analysis and artificial neural network. *J Mater Process Technol*. 2005;170:494−500. https://doi.org/10.1016/j.jmatprotec.2005.04.072.
21. Wu Y, Yuan M, Dong S, Lin L, Liu Y. Remaining useful life estimation of engineered systems using vanilla LSTM neural networks. *Neurocomputing*. 2018;275:167−179. https://doi.org/10.1016/j.neucom.2017.05.063.
22. Tian Z. An artificial neural network method for remaining useful life prediction of equipment subject to condition monitoring. *J Intell Manuf*. 2012;23:227−237. https://doi.org/10.1007/s10845-009-0356-9.
23. Tian Z, Wong L, Safaei N. A neural network approach for remaining useful life prediction utilizing both failure and suspension histories. *Mech Syst Signal Process*. 2010;24:1542−1555. https://doi.org/10.1016/j.ymssp.2009.11.005.
24. Shao H, Jiang H, Zhang X, Niu M. Rolling bearing fault diagnosis using an optimization deep belief network. *Meas Sci Technol*. 2015;26:115002. https://doi.org/10.1088/0957-0233/26/11/115002.
25. Mainorwirth, https://mainorwirth.com/blog/autopilot-autonomous-differences/ [as accessed on 10 July 2021].
26. S. Kuutti, https://deepai.org/publication/a-survey-of-deep-learning-applications-to-autonomous-vehicle-control.
27. Kuutti S, Bowden R, Jin Y, Barber P, Fallah S. A survey of deep learning applications to autonomous vehicle control. *IEEE Trans Intell Transport Syst*. 2021;22:712−733. https://doi.org/10.1109/TITS.2019.2962338.
28. J. Cohen, https://thinkautonomous.medium.com/ [as accessed on 114 July 2021].

29. Audi, https://www.audi-mediacenter.com/en/audi-ai-9099/audi-ai-the-technology-9101 [as accessed on 1 June 2021].
30. Geng K, Yin G. Using deep learning in infrared images to enable human gesture recognition for autonomous vehicles. *IEEE Access*. 2020;8:88227–88240. https://doi.org/10.1109/ACCESS.2020.2990636.
31. Walz, https://www.futurecar.com/3929/Waymo-Develops-a-Machine-Learning-Model-to-Predict-the-Behavior-of-Other-Road-Users-for-its-Self-Driving-Vehicles [as accessed on 14 June 2021].
32. Zhang E, Xue B, Cao F, Duan J, Lin G, Lei Y. Fusion of 2d cnn and 3d densenet for dynamic gesture recognition. *Electronics*. 2019;8:1511. https://doi.org/10.3390/electronics8121511.
33. Liu H, Deng Z, Xu C. Learning spatiotemporal features with 3D DenseNet and attention for gesture recognition. *Int J Electr Eng Educ*. 2019. https://doi.org/10.1177/0020720919894196, 0020720919894196.
34. Wan L, Sun Y, Sun L, Ning Z, Rodrigues JJ. Deep learning based autonomous vehicle super resolution DOA estimation for safety driving. *IEEE Trans Intell Transport Syst*. 2021;22. https://doi.org/10.1109/TITS.2020.3009223.
35. Wang H, Wan L, Dong M, Ota K, Wang X. Assistant vehicle localization based on three collaborative base stations via SBL-based robust DOA estimation. *IEEE Internet Things J*. 2019;6:5766–5777. https://doi.org/10.1109/JIOT.2019.2905788.
36. Ibrahim MS, Dong W, Yang Q. Machine learning driven smart electric power systems: current trends and new perspectives. *Appl Energy*. 2020;272:115237. https://doi.org/10.1016/j.apenergy.2020.115237.
37. Smartgrid.gov, https://www.smartgrid.gov/the_smart_grid/smart_grid.html [as accessed on 2 July 2021].
38. Ozay M, Esnaola I, Vural FTY, Kulkarni SR, Poor HV. Machine learning methods for attack detection in the smart grid. *IEEE Transact Neural Network Learn Syst*. 2015;27:1773–1786. https://doi.org/10.1109/TNNLS.2015.2404803.
39. Yu W, An D, Griffith D, Yang Q, Xu G. Towards statistical modeling and machine learning based energy usage forecasting in smart grid. *ACM SIGAPP - Appl Comput Rev*. 2015;15:6–16. https://doi.org/10.1145/2753060.2753061.
40. Esmalifalak M, Liu L, Nguyen N, Zheng R, Han Z. Detecting stealthy false data injection using machine learning in smart grid. *IEEE Syst J*. 2014;11:1644–1652. https://doi.org/10.1109/JSYST.2014.2341597.
41. Jindal A, Dua A, Kaur K, Singh M, Kumar N, Mishra S. Decision tree and SVM-based data analytics for theft detection in smart grid. *IEEE Trans Ind Inf.* 2016;12:1005–1016. https://doi.org/10.1109/TII.2016.2543145.
42. Jiao W, Li VO. Support vector machine detection of data framing attack in smart grid. In: *Proceedings of IEEE Conference on Communications and Network Security (CNS)*. Beijing, China: IEEE; 30 May–1 June, 2018:1–5. https://doi.org/10.1109/CNS.2018.8433210.
43. Kaygusuz C, Babun L, Aksu H, Uluagac AS. Detection of compromised smart grid devices with machine learning and convolution techniques. In: *Proceedings of IEEE International Conference on Communications (ICC), Kansas City, MO, USA*. 20–24 May, 2018:1–6. https://doi.org/10.1109/ICC.2018.8423022.
44. Ali AS, Azad S. *Demand Forecasting in Smart Grid, Smart Grids*. Springer; 2013:135–150. https://doi.org/10.1007/978-1-4471-5210-1_6.
45. Zhou K, Yang S-l, Shen C. *A Review of Electric Load Classification in Smart Grid Environment, Renewable and Sustainable Energy Reviews*. Vol. 24. Elsevier; 2013:103–110. https://doi.org/10.1016/j.rser.2013.03.023.

46. Lu Y, Zhang T, Zeng Z. Adaptive weighted fuzzy clustering algorithm for load profiling of smart grid customers. In: *Proceedings of IEEE/CIC International Conference on Communications in China (ICCC), Chengdu, China*. 27–29 July, 2016:1–6. https://doi.org/10.1109/ICCChina.2016.7636874.
47. University of Tartu, https://sisu.ut.ee/imageprocessing/book/1 [as accessed on 15 June 2021].
48. Edx.org, https://www.edx.org/learn/mechanical-engineering [as accessed on 5 June 2021].
49. Times higher education, https://www.timeshighereducation.com/student/subjects/what-can-you-do-electrical-engineering-degree [as accessed on 30 March 2021].
50. Aslam Y, Santhi N, Ramasamy N, Ramar K. Localization and segmentation of metal cracks using deep learning. *J Ambient Intell Hum Comput*. 2021;12:4205–4213. https://doi.org/10.1007/s12652-020-01803-8.
51. Choi W, Huh H, Tama BA, Park G, Lee S. A neural network model for material degradation detection and diagnosis using microscopic images. *IEEE Access*. 2019;7: 92151–92160. https://doi.org/10.1109/ACCESS.2019.2927162.
52. Wei X, Jiang S, Li Y, Li C, Jia L, Li Y. Defect detection of pantograph slide based on deep learning and image processing technology. *IEEE Trans Intell Transport Syst*. 2019; 21:947–958. https://doi.org/10.1109/TITS.2019.2900385.
53. Jiang S, Wei X, Yang Z. Defect detection of pantograph slider based on improved faster R-CNN. In: *Proceedings of Chinese Control and Decision Conference (CCDC), Nanchang, China*. 3–5 June, 2019:5278–5283. https://doi.org/10.1109/CCDC.2019.8832401.
54. Shanmugamani R, Sadique M, Ramamoorthy B. Detection and classification of surface defects of gun barrels using computer vision and machine learning. *Measurement*. 2015;60:222–230. https://doi.org/10.1016/j.measurement.2014.10.009.
55. Sugumaran V, Ramachandran K. Effect of number of features on classification of roller bearing faults using SVM and PSVM. *Expert Syst Appl*. 2011;38:4088–4096. https://doi.org/10.1016/j.eswa.2010.09.072.
56. Sugumaran V, Muralidharan V, Ramachandran K. Feature selection using decision tree and classification through proximal support vector machine for fault diagnostics of roller bearing. *Mech Syst Signal Process*. 2007;21:930–942. https://doi.org/10.1016/j.ymssp.2006.05.004.
57. Tan C, Lv S, Dong F, Takei M. Image reconstruction based on convolutional neural network for electrical resistance tomography. *IEEE Sensor J*. 2018;19:196–204. https://doi.org/10.1109/JSEN.2018.2876411.
58. Li L, Ota K, Dong M. Everything is image: CNN-based short-term electrical load forecasting for smart grid. In: *Proceedings of 14th International Symposium on Pervasive Systems, Algorithms and Networks & 2017 11th International Conference on Frontier of Computer Science and Technology & 2017 Third International Symposium of Creative Computing (ISPAN-FCST-ISCC)*. IEEE; 21–23 June, 2017:344–351. https://doi.org/10.1109/ISPAN-FCST-ISCC.2017.78.
59. Lai J-P, Chang Y-M, Chen C-H, Pai P-F. A survey of machine learning models in renewable energy predictions. *Appl Sci*. 2020;10:5975. https://doi.org/10.3390/app10175975.
60. Wang H, Lei Z, Zhang X, Zhou B, Peng J. A review of deep learning for renewable energy forecasting. *Energy Convers Manag*. 2019;198:111799. https://doi.org/10.1016/j.enconman.2019.111799.
61. Demolli H, Dokuz AS, Ecemis A, Gokcek M. Wind power forecasting based on daily wind speed data using machine learning algorithms. *Energy Convers Manag*. 2019;198: 111823. https://doi.org/10.1016/j.enconman.2019.111823.

62. Zucatelli P, Nascimento E, Aylas G, et al. Short-term wind speed forecasting in Uruguay using computational intelligence. *Heliyon*. 2019;5:e01664.
63. Innani A, Dhatwalia S, Tripathi Y, jambhulkar V. Solar power predictor using ensemble learning. *Int Res J Eng Technol*. 2020;7.
64. Shayan E, Zare V, Mirzaee I. Hydrogen production from biomass gasification; a theoretical comparison of using different gasification agents. *Energy Convers Manag*. 2018; 159:30–41. https://doi.org/10.1016/j.enconman.2017.12.096.
65. Elmaz F, Yücel Ö, Mutlu AY. *Predictive Modeling of Biomass Gasification with Machine Learning-Based Regression Methods, Energy*. Vol. 191. Elsevier; 2020:116541. https://doi.org/10.1016/j.energy.2019.116541.
66. Kavousi-Fard A, Su W. A combined prognostic model based on machine learning for tidal current prediction. *IEEE Trans Geosci Rem Sens*. 2017;55:3108–3114. https://doi.org/10.1109/TGRS.2017.2659538.
67. Riazi A. Accurate tide level estimation: a deep learning approach. *Ocean Eng*. 2020; 198:107013. https://doi.org/10.1016/j.oceaneng.2020.107013.
68. Chen L, Lai X. Comparison between ARIMA and ANN models used in short-term wind speed forecasting. In: *Proceedings of Asia-Pacific Power and Energy Engineering Conference, Wuhan, China*. 25–28 March, 2011:1–4. https://doi.org/10.1109/APPEEC.2011.5748446.
69. Brahimi T. Using artificial intelligence to predict wind speed for energy application in Saudi Arabia. *Energies*. 2019;12:4669. https://doi.org/10.3390/en12244669.
70. Alexiadis M, Dokopoulos P, Sahsamanoglou H, Manousaridis I. Short-term forecasting of wind speed and related electrical power. *Sol Energy*. 1998;63:61–68. https://doi.org/10.1016/S0038-092X(98)00032-2.
71. Soman SS, Zareipour H, Malik O, Mandal P. *A Review of Wind Power and Wind Speed Forecasting Methods with Different Time Horizons, North American Power Symposium 2010*. Arlington, TX, USA: IEEE; 2010:1–8. https://doi.org/10.1109/NAPS.2010.5619586, 26-28 Sept. 2010.
72. Liu H, Tian H-q, Pan D-f, Li Y-f. Forecasting models for wind speed using wavelet, wavelet packet, time series and Artificial Neural Networks. *Appl Energy*. 2013;107: 191–208. https://doi.org/10.1016/j.apenergy.2013.02.002. Elsevier.
73. Bilgili M, Sahin B, Yasar A. Application of artificial neural networks for the wind speed prediction of target station using reference stations data. *Renew Energy*. 2007; 32:2350–2360. https://doi.org/10.1016/j.renene.2006.12.001.
74. Mohandes MA, Halawani TO, Rehman S, Hussain AA. Support vector machines for wind speed prediction. *Renew Energy*. 2004;29:939–947. https://doi.org/10.1016/j.renene.2003.11.009.
75. Gangwar S, Bali V, Kumar A. Comparative analysis of wind speed forecasting using LSTM and SVM. *EAI Endors Transact Scal Informat Syst*. 2020;7. https://doi.org/10.4108/eai.13-7-2018.159407.
76. Zhou J, Shi J, Li G. Fine tuning support vector machines for short-term wind speed forecasting. *Energy Convers Manag*. 2011;52:1990–1998. https://doi.org/10.1016/j.enconman.2010.11.007.
77. Salcedo-Sanz S, Ortiz-Garcı EG, Pérez-Bellido ÁM, Portilla-Figueras A, Prieto L. Short term wind speed prediction based on evolutionary support vector regression algorithms. *Expert Syst Appl*. 2011;38:4052–4057. https://doi.org/10.1016/j.eswa.2010.09.067.
78. Lahouar A, Slama JBH. Wind speed and direction prediction for wind farms using support vector regression, 2014 5th International Renewable Energy Congress (IREC). *IEEE, Hammamet, Tunisia*. 25–27 March, 2014:1–6. https://doi.org/10.1109/IREC.2014.6826932.

79. Osama S, Darwish A, Houssein EH, Hassanien AE, Fahmy AA, Mahrous A. Long-term wind speed prediction based on optimized support vector regression. In: *Proceedings of Eighth International Conference on Intelligent Computing and Information Systems (ICICIS)*. Cairo, Egypt: IEEE; 5–7 December, 2017:191–196. https://doi.org/10.1109/INTELCIS.2017.8260035.
80. Santamaría-Bonfil G, Reyes-Ballesteros A, Gershenson C. Wind speed forecasting for wind farms: a method based on support vector regression. *Renew Energy*. 2016;85:790–809. https://doi.org/10.1016/j.renene.2015.07.004.
81. Voyant C, Notton G, Kalogirou S, et al. Machine learning methods for solar radiation forecasting: a review. *Renew Energy*. 2017;105:569–582. https://doi.org/10.1016/j.renene.2016.12.095.
82. Şenkal O, Kuleli T. Estimation of solar radiation over Turkey using artificial neural network and satellite data. *Appl Energy*. 2009;86:1222–1228. https://doi.org/10.1016/j.apenergy.2008.06.003.
83. Yadav AK, Chandel S. Solar radiation prediction using artificial neural network techniques: a review. *Renew Sustain Energy Rev*. 2014;33:772–781. https://doi.org/10.1016/j.rser.2013.08.055.
84. Sasi T, Mighani M, Örs E, Tawani R, Gräbner M. Prediction of ash fusion behavior from coal ash composition for entrained-flow gasification. *Fuel Process Technol*. 2018;176:64–75. https://doi.org/10.1016/j.fuproc.2018.03.018.
85. Dahikar SS, Rode SV. Agricultural crop yield prediction using artificial neural network approach. *Int J Innovat Res Electric, Electron, Instrument Control Eng*. 2014;2:683–686.
86. Guo WW, Xue H. Crop yield forecasting using artificial neural networks: a comparison between spatial and temporal models. *Math Probl Eng*. 2014;2014:857865. https://doi.org/10.1155/2014/857865.
87. Gandhi N, Petkar O, Armstrong LJ. Rice crop yield prediction using artificial neural networks. In: *Proceedings of IEEE Technological Innovations in ICT for Agriculture and Rural Development (TIAR), Chennai, India*. 15–16 July, 2016:105–110. https://doi.org/10.1109/TIAR.2016.7801222.
88. Li L, Yuan Z, Gao Y. Maximization of energy absorption for a wave energy converter using the deep machine learning. *Energy*. 2018;165:340–349. https://doi.org/10.1016/j.energy.2018.09.093.
89. Zhao Y-P, Bi C-W, Sun X-X, Dong G-H. A prediction on structural stress and deformation of fish cage in waves using machine-learning method. *Aquacult Eng*. 2019;85:15–21. https://doi.org/10.1016/j.aquaeng.2019.01.003.
90. Chang H-K, Liou J-C, Liu S-J, Liaw S-R. Simulated wave-driven ANN model for typhoon waves. *Adv Eng Softw*. 2011;42:25–34. https://doi.org/10.1016/j.advengsoft.2010.10.014.
91. Kavousi-Fard A. Modeling uncertainty in tidal current forecast using prediction interval-based SVR. *IEEE Trans Sustain Energy*. 2016;8:708–715. https://doi.org/10.1109/TSTE.2016.2606488.
92. Guozhen Y, Haifeng W, Hui Q, Jianming F. *Tidal Current Short-Term Prediction Based on Support Vector Regression, IOP Conference Series: Materials Science and Engineering*. Vol. 199. IOP Publishing; 2017:012024. https://doi.org/10.1088/1757-899x/199/1/012024.
93. Rasel RI, Uddin MN, Islam F, Haroon A. Application of deep neural network for predicting river tide level. In: *Proceedings of International Conference on Innovations in Science, Engineering and Technology (ICISET)*. Chittagong, Bangladesh: IEEE; 27–28 October, 2018:311–316. https://doi.org/10.1109/ICISET.2018.8745593.

94. Kabir IFSA, Ng EYK. Effect of different atmospheric boundary layers on the wake characteristics of NREL phase VI wind turbine. *Energy*. 2019;130:1185–1197. https://doi.org/10.1016/j.renene.2018.08.083.
95. Kabir IFSA, Ng EYK. Insight into stall delay and computation of 3D sectional aerofoil characteristics of NREL phase VI wind turbine using inverse BEM and improvement in BEM analysis accounting for stall delay effect. *Energy*. 2017;120:518–536. https://doi.org/10.1016/j.energy.2016.11.102.
96. Purohit S, Kabir IFSA, Ng EYK. On the accuracy of uRANS and LES-based CFD modeling approaches for rotor and wake aerodynamics of the (New) Mexico wind turbine rotor Phase-III. *Energies*. 2021;14:5198. https://doi.org/10.3390/en14165198. special issue in Advancement in Wind Turbine Technology (26 pages).
97. Kabir IFSA, Safiyullah F, Ng EYK, Tam VWY. New analytical wake models based on artificial intelligence and rivalling the benchmark full-rotor CFD predictions under both uniform and ABL inflows. *Energy*. 2020;193:116761. https://doi.org/10.1016/j.energy.2019.116761.
98. Purohit S, Ng EYK, Kabir IFSA. Evaluation of three potential machine learning algorithms for predicting the velocity and turbulence intensity of a wind turbine wake. *Renew Energy*. 2021;184:405–420. https://doi.org/10.1016/j.renene.2021.11.097.
99. Hasan F, Kargarian A, Mohammadi A. A survey on applications of machine learning for optimal power flow. In: *Proceedings of IEEE Texas Power and Energy Conference (TPEC), College Station, TX, USA*. 6–7 February, 2020:1–6. https://doi.org/10.1109/TPEC48276.2020.9042547.
100. Baker K. Learning warm-start points for AC optimal power flow. In: *IEEE 29th International Workshop on Machine Learning for Signal Processing (MLSP), Pittsburgh, PA, USA*. 13–16 October, 2019:1–6. https://doi.org/10.1109/MLSP.2019.8918690.
101. Owerko D, Gama F, Ribeiro A. Optimal power flow using graph neural networks. In: *Proceedings of ICASSP IEEE International Conference on Acoustics, Speech and Signal Processing (ICASSP), Barcelona, Spain*. 4–8 May, 2020:5930–5934. https://doi.org/10.1109/ICASSP40776.2020.9053140.
102. Pan X, Zhao T, Chen M. DeepOPF: deep neural network for DC optimal power flow. In: *Proceedings of IEEE International Conference on Communications, Control, and Computing Technologies for Smart Grids (SmartGridComm), Beijing, China*. 21–23 October, 2019:1–6. https://doi.org/10.1109/SmartGridComm.2019.8909795.
103. Zhao T, Pan X, Chen M, Venzke A, Low SH. DeepOPF+: a deep neural network approach for DC optimal power flow for ensuring feasibility. In: *Proceedings of IEEE International Conference on Communications, Control, and Computing Technologies for Smart Grids (SmartGridComm), Tempe, AZ, USA*. 11–13 November, 2020:1–6. https://doi.org/10.1109/SmartGridComm47815.2020.9303017.
104. Osman M, Abo-Sinna MA, Mousa A. A solution to the optimal power flow using genetic algorithm. *Appl Math Comput*. 2004;155:391–405. https://doi.org/10.1016/S0096-3003(03)00785-9.
105. Singh SP, Tyagi R, Goel A. Genetic algorithm for solving the economic load dispatch. *Int J Electron Electr Eng*. 2014;7, 523-523.
106. Nanda J, Narayanan RB. Application of genetic algorithm to economic load dispatch with Lineflow constraints. *Int J Electr Power Energy Syst*. 2002;24:723–729. https://doi.org/10.1016/S0142-0615(02)00002-9.
107. Baker K, Bernstein A. Joint chance constraints reduction through learning in active distribution networks. In: *Proceedings of IEEE Global Conference on Signal and Information Processing (GlobalSIP), Anaheim, CA, USA*. 26–29 November, 2018:922–926. https://doi.org/10.1109/GlobalSIP.2018.8646440.
108. Trendmicro, https://www.trendmicro.com/vinfo/in/security/definition/industrial-control-system [as accessed on 15 July 2021].

109. Keliris A, Salehghaffari H, Cairl B, Krishnamurthy P, Maniatakos M, Khorrami F. Machine learning-based defense against process-aware attacks on industrial control systems. In: *Proceedings of IEEE International Test Conference (ITC), Fort Worth, TX, USA*. 15–17 November, 2016:1–10. https://doi.org/10.1109/TEST.2016.7805855.
110. Al-Abassi A, Karimipour H, Dehghantanha A, Parizi RM. An ensemble deep learning-based cyber-attack detection in industrial control system. *IEEE Access*. 2020;8:83965–83973. https://doi.org/10.1109/ACCESS.2020.2992249.
111. Kravchik M, Shabtai A. Detecting cyber attacks in industrial control systems using convolutional neural networks. In: *Proceedings of the 2018 Workshop on Cyber-Physical Systems Security and PrivaCy*. 2018:72–83. https://doi.org/10.1145/3264888.3264896.
112. Sokolov AN, Pyatnitsky IA, Alabugin SK. Research of classical machine learning methods and deep learning models effectiveness in detecting anomalies of industrial control system. In: *Proceedings of Global Smart Industry Conference (GloSIC)*. Chelyabinsk, Russia: IEEE; 13–15 November, 2018:1–6. https://doi.org/10.1109/GloSIC.2018.8570073.
113. Yasami Y, Mozaffari SP. A novel unsupervised classification approach for network anomaly detection by k-Means clustering and ID3 decision tree learning methods. *J Supercomput*. 2010;53:231–245. https://doi.org/10.1007/s11227-009-0338-x.
114. Hijazi A, El Safadi A, Flaus J-M. *A Deep Learning Approach for Intrusion Detection System in Industry Network, BDCSIntell*. 2018:55–62.
115. Amato F, Mazzocca N, Moscato F, Vivenzio E. Multilayer perceptron: an intelligent model for classification and intrusion detection. In: *Proceedings of 31st International Conference on Advanced Information Networking and Applications Workshops (WAINA)*. Taipei, Taiwan: IEEE; 27–29 March, 2017:686–691. https://doi.org/10.1109/WAINA.2017.134.
116. Dongqing C, Puhan Z, Huazhong W. Intrusion detection for industrial control systems based on an improved SVM method. *J Tsinghua Univ (Sci Technol)*. 2018;58:380–386. https://doi.org/10.16511/j.cnki.qhdxxb.2018.25.019.
117. Khorrami F, Krishnamurthy P, Karri R. Cybersecurity for control systems: a process-aware perspective. *IEEE Design Test*. 2016;33:75–83. https://doi.org/10.1109/MDAT.2016.2594178.
118. Zeeshan-ul-hassan Usmani, https://www.kaggle.com/getting-started/44916 [as accessed on 5 June 2021].
119. Google search central, https://developers.google.com/search/docs/advanced/structured-data/dataset [as accessed on 4 July 2021].

CHAPTER TWO

A Multiple Object Recognition Approach via DenseNet-161 Model

Anamika Dhillon and Gyanendra K. Verma
Department of Computer Engineering, National Institute of Technology, Kurukshetra, India

Contents

1. Introduction

The main purpose of object detection is to identify and locate one or more targets from a particular image or video dataset. It has large application prospects in various areas, such as road traffic accident prevention, wild

Smart Electrical and Mechanical Systems
ISBN: 978-0-323-90789-7
https://doi.org/10.1016/B978-0-323-90789-7.00009-9

animal detection, face recognition, etc. The scenarios for detecting the multitarget objects are usually complicated and variable due to several reasons like low-resolution images, where the images are not sharply defined, variation in pose, blur, occlusion, etc., as shown in Fig. 2.1.

Currently, deep learning[1–3] has become the most popular solution for scene image classification and detection. It performs differently than the classical approaches, as they are the powerful framework for learning the representation of features from the data. DCNNs (deep convolutional neural networks)[1,4] have become a presiding approach in machine learning and their notable trend is that their models continue to go deeper. From AlexNet to GoogLeNet and the ResNet, the accuracy and depth of convolutional neural network (CNN)[5–7] models increased with the increasing number of layers.

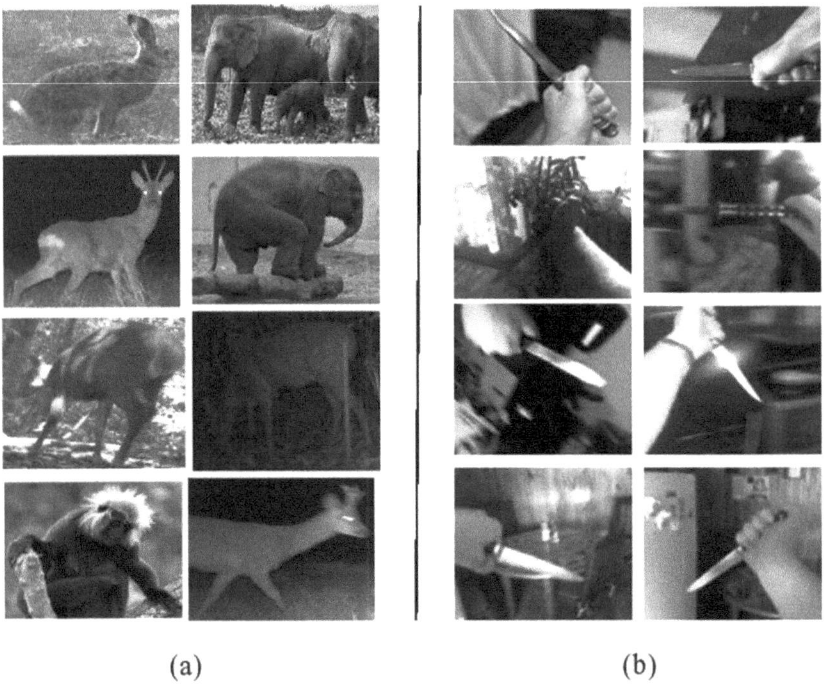

Figure 2.1 Sample images of the datasets: (A) wild animal and (B) knife. *From Zhang Z, He Z, Cao G, Cao W. Animal detection from highly cluttered natural scenes using spatio-temporal object region proposals and patch verification.* IEEE Trans Multimedia. *2016; 18(10):2079–2092. https://doi.org/10.1109/TMM.2016.2594138.*

In order to achieve high-level feature maps, we have utilized DenseNet-161 model, directly to our database. DenseNet has various compelling advantages: (1) strong gradient flow: The signals with error can be effortlessly propagated to previous layers more straightforwardly; (2) parameter and computational efficiency: A number of parameters in DenseNet are proportionate to $1 \times k \times k$, rather than $C \times C$; since $k << C$, DenseNet is smaller in size than ResNet and also supports the reuse of features and reduction in the number of parameters; (3) more heterogeneous and low complexity features; and (4) mitigating the vanishing gradient problem. DenseNet has some drawbacks also: (1) For L-layer DenseNet, there exist a total of $L(L-1)/2$ connections. With the increase in number of connections, network's computations efficiency decreases. (2) The correlation among interlayer feature map is not expressly modeled. In this chapter, we first introduced animal and knife datasets for detecting the target images. The database is distributed into two parts, positive images and negative images. Secondly, we introduce DenseNet-161 to classify the given images. We comparatively evaluate our algorithm against BCNet, ResNet, and CNN. The major contribution of this chapter can be summarized as follows:

- We have evolved a systematic network for the recognition of various objects including animals and knives.
- We have found that even in the highly cluttered images, including camera trap images, the DenseNet-161 model classifies the objects more accurately from each other.
- We present an evaluation, comparison, and abstract survey on various purposes with two different classes, i.e., positive and negative.

We have organized the chapter as follows. Section 2 explains the related work for the detection of objects, including animal and knife datasets. Section 3 specifies the overview of DenseNet architecture; Section 4 explains the methodology for propounded work. Section 5 describes the result analysis and discussion, and Section 6 provides the conclusion of the chapter.

2. Related work

This section presents some of the work closely related to DenseNet-161 and its applications in various areas of image processing.

2.1 DenseNet

Gao Huang[8] proposed a technique known as densely connected convolutional networks for detecting and recognizing objects. This chapter collects the superlative information by connecting each layer to every other layer. They have utilized various datasets namely, CIFAR, SVHN, and ImageNet.[9] By increasing the number of parameters, they have attained a TOP-5 error rate of 5.17%. Ke Zhang et al.[10] proposed a framework for the classification of an image by using multiple features reweight DenseNet (MFR-DenseNet). An error rate of 3.57% and 18.27% has been attained on CIFAR-10 and CIFAR-100, respectively. Ke Zhang et al.[11] propounded the image classification technique by utilizing channel-wise and feature points reweights DenseNet model. The experiments were performed on CIFAR-10/100 datasets and the results are outperformed by 5.6% with fewer parameters. Guandong Li et al. proposed an image classification method by utilizing an embedded squeeze-and-excitation (SE)[12] technique with 3D DenseNet. The SE block enhances the interdependencies among the channels and extracts the essential knowledge from the convolutional layers. The experimentations have been accomplished on Indian Pines and Pavia University HSI databases and attained an accuracy of 99.95% and 99.96%, respectively. Sheping Zhai et al.[13] proposed an object detection method by utilizing an improved version of the SSD algorithm.[14] They have propounded a framework known as DenseNet-S-32-1, which immersed the subjective information with pyramid feature layers. They have achieved an efficiency of 81.4%, 79.0%, and 29.5% on PASCAL VOC 2007/2012 and MS COCO datasets. The notable reviewed models of DenseNet-161 are listed in Table 2.1.

2.2 Spatiotemporal distilled dense-connectivity network

Wangli Hao (2019)[15] proposed a framework for the action recognition known as spatiotemporal distilled dense-connectivity network. DenseNet stimulates their model by comprising of two block-level subnetworks which are densely connected and one knowledge edge distillation component. The experiments are done on HMDB51 and UCF101 datasets and have attained an accuracy of 69.49% and 94.79% on these respective datasets.

2.3 Efficient ladder-style DenseNets

Ivan Kreso[16] proposed is a DenseNet-based ladder-style framework that has been utilized for semantic segmentation. This method consists of two data paths: the first one is downsampling data path and the second one is upsampling data path. The downsampling data path comprises of DenseNet Feature extractor and lightweight spatial pyramid pooling module.[17] They have achieved 80.6% IoU on city spaces and 83.6% IoU on PASCAL VOC 2012 datasets.

Table 2.1 Noticeable papers for DenseNet.

Author	Year	Methods	Description	Dataset
G. Huang et al.[8]	2018	DenseNet	Object recognition	CIFAR-10, CIFAR-100 ImageNet
K. Zhang et al.[10]	2018	MFR-DenseNet	Image classification	CIFAR-10, CIFAR-100
K. Zhang et al.[11]	2019	CAPR-DenseNet	Image classification	CIFAR-10, CIFAR-100
S. Zhai et al.[13]	2020	DF-SSD	Object detection	PASCAL VOC 2007 and 2012
W. Hao et al.[15]	2019	DenseNet	Video action recognition	UCF101HMDB51
I. Kreso et al.[16]	2019	Efficient ladder-style DenseNets	Semantic segmentation	Cityscapes, PASCAL VOC 2012 and ROB 2018
T. Tong et al.[18]	2017	Dense skip connections	Image super-resolution	Set5, Set14, Urban100
W. Wang et al.[19]	2019	Models DenseNet-MobileNet	Image classification	Caltech-101, Uebingen animals
P. Shamsolmoali et al.[21]	2019	Dilated dense progressive network	Image super-resolution	Set5, Urban100, B100

2.4 Dense skip connections

Tong Tong[18] propounded a super-resolution DenseNet method, which is disintegrated into distinct parts, such as a block of DenseNet, deconvolutional layers, and the reconstruction layers. The performance of this framework is evaluated on four datasets, namely, Set5, Set14, B100, and Urban100.

2.5 DenseNet-MobileNet models

Wei Wang[19] proposed a Dense-MobileNet model for the classification of images. In this model, the convolutional layer, which is similar in size of the input feature maps in the MobileNet model,[20] are taken as the dense blocks, within which dense connections are made. They have designed two different Dense-MobileNet Models: the first one is Dense1-MobileNet and the second one is Dense2-MobileNet. They have done the experiments on Caltech-101 and Uebingen Animals datasets. On the Caltech-101 dataset, both the models attain classification accuracy of 76.6% and 77.67%, respectively, and on Uebingen Animals dataset, they achieve 90.65% and 92.1% accuracy.

2.6 Dilated dense progressive network

Pourya Shamsolmoali[21] proposed a method where they have combined the dilated dense convolutional network with the residual network. In this model, except for the layer which is utilized for local as well as global feature integration, all other layers are of size 3 × 3. The efficiency of this model is evaluated on five datasets, namely, Set4, Set14, BSD100, Urban100, and a few real-world datasets.

3. DenseNet architecture

DenseNet was propounded by Gao et al. in 2017 and its basic unit is dense block. We illustrate the efficiency and performance of our dataset on the DenseNet-161. It consists of densely connected CNN layers to ensure the maximum flow of information. DenseNet is devised with dense connectivity[22] among its layer, and this concept is adequate for feature reuse.

The feature map of the L-1 layer is demonstrated as s × s × k_0, where s stands for feature map's size and k_0 stands for the number of channels. A nonlinear transformation is represented by H(·), where a sequence of operations[4] are incorporated, such as batch normalization (BN), rectified linear unit (ReLU), a convolutional layer of 1 × 1 and 3 × 3, as represented in Fig. 2.2.

Figure 2.2 Architecture of DenseNet. *From Huang G, Liu Z, Van Der Maaten L, Weinberger KQ. Densely connected convolutional networks. In:* Proceedings - 30th IEEE Conference on Computer Vision and Pattern Recognition, *CVPR 2017, Institute of Electrical and Electronics Engineers Inc. vols 2017: 2261–2269. https://doi.org/10.1109/CVPR.2017.243.*

To minimize the number of channels, 1 × 1 convolution operation is incorporated, whereas for reorganization of the features, 3 × 3 convolution is utilized; H(·) alters the number of channels to k (k = 32). Due to the dense interconnections, feature map of L-1 layer is immediately coupled with the L layer, then concatenates with H(·)'s output, and thus represents the outcome as $s \times s \times (k_0 + k)$. Likewise, $s \times s \times (k_0 + 2k)$ is the outcome of L + 1 layer. The transition layer consists of the following three layers: BN, convolutional layer, and the pooling layer. DenseNet's deeper layers depend on the high-level features, which is further responsible for the enhancement of growth rate.

3.1 DenseNet layers

DenseNet architecture contains dense convolutional blocks along with straight interconnections from a particular layer to every following layer so that the flow of data on the input images of size 224 × 224 can be enhanced. The layer which exists between the two adjacent dense blocks is indicated as the transition layer, and it is comprised of 1 × 1 convolution and 2 × 2 average pooling. At each side of the input, one-pixel padding is implied for 3 × 3 convolutional layers. For alleviating the number of input feature maps and upgrading the computational efficiency, the bottleneck layer is utilized before each 3 × 3 convolution. At dense blocks, a softmax classifier and global average pooling is applied.

3.2 Efficiency of DenseNet

The major difference among the DenseNet and the already existing frameworks is that DenseNet can have extreme narrow layers, e.g., k = 12, where k is the growth rate. It implies that every layer in the Dense block will process k features, which will be further concatenated with the framework. DenseNet has remarkably lower validation error with the same amount of features.

Remarks

- Concatenating feature maps enhances the distinction in input of succeeding layers and improves the efficiency.
- Minimize the problem of vanishing gradient.
- Support feature reuse.
- Decrease the number of parameters.

4. Methodology

4.1 Background

In DenseNet, all feature maps are concatenated to assure the maximum flow of information. If the output of the $(l-1)^{th}$ layer is x_l-1, then the output of the lth layer is

$$x_l = H_l([x_0, x_1 .... x_l - 1])$$

where H(·) executes a sequence of consecutive transformation, i.e., BN, ReLU, and then convolution.

4.2 Evaluation metrics

Following metrics are utilized to evaluate the performance: accuracy, error rate, and the confusion matrix. An accuracy[23] metric is utilized to calculate the performance of model in an explicable manner and it is evaluated in percentage form. Determination of the model's accuracy has been done besides the parameters of the model. When the binary classifications are performed, then accuracy can be determined in positive and negative terms, as follows:

$$\mathrm{TPR} = \frac{\mathrm{TP} + \mathrm{TN}}{\mathrm{TP} + \mathrm{FP} + \mathrm{FN} + \mathrm{TN}}$$

True positive means the positive classes are accurately predicted, whereas the true negative describes that the negative classes are precisely estimated. When the positive classes are predicted inaccurately, it results in the false-positive value, while false negative describes that the negative class is not correctly identified. Training a model is a process of imbibing good values for the weights and the bias from labeled examples. Inadequate prediction will lead to loss of results and this loss is numeric data, which defines the inefficiency in model's prediction on an example. Model's prediction perfection is indirectly proportionate to the loss. The aim of training a model is to strike the balance of weights and biases with low loss, on average, throughout all across. Higher loss indicates bad prediction in a framework. For any model, the interpretation of training and validation, calculate the loss for these two sets, where the accuracy is calculated in terms of prediction, wherein the loss is a sum of error, made for every example during training or validation set. In the domain of machine learning, a confusion matrix is also called an error matrix, which is a designed tabular representation for the performance of algorithm. The name confusion matrix itself reveals the fact that it enables easy identification of confusion between the classes or commonly mislabeled classes one as another. The confusion matrix is used to compute the most of the performance measures.

4.3 Dataset

In this chapter, we have used DenseNet for two applications, which are as follows: wild animals and knife detection.

4.3.1 Wild animal dataset

Several latest wild animal image database are assembled a short while ago and they are easily accessible to public. Karina Figueroa et al.[24] propounded a fast technique for the detection of wild animals captured from the camera trap images. They have utilized two procedures: the first one is called intuitive technique, in which pixel difference of two images has been taken, and the second one is utilized to perceive the regions of different patterns. This method is being used for identifying the ocelots of an image. Marcin Pietrasik[25] proposed a technique for detecting animals in man-made region. Jason Parham et al.[26] proposed a pipeline technique which is used to recognize the animals. They have introduced a five-stage detection method, which includes selection of input image, classification of image, annotation localization, annotation classification, background segmentation, and finally annotations of interest classification. They have used WILD dataset and obtained a localization mAP of 81.67%. Alexander Loos et al.[27] proposed a framework for wild animal's classification and localization. To increase the performance of their technique, they have used you only look once (YOLO)[28] and single shot detector (SSD) as detectors. They have performed the experiments on Snapshot Serengeti dataset and accomplished a rate of accurately calculated animals as 82%. William H. S. Antonio et al.[29] proposed an animal detection framework by utilizing machine learning methods. The features of two categories, namely animals and no animals, are classified and extracted in their work. They have used two perspectives, i.e., K-nearest neighbors and random forest, and got the best F-measure of 0.6243. Benjamin Kellenberger et al.[30] proposed a framework for detecting the animals by utilizing DCNN and active learning. This method was performed on unmanned aerial vehicle images, and for this, they adopt a new approach called transfer sampling. Window cropping is utilized to obtain additional labels per question. By using screening, they capture 80% animals from the dataset. Gyanendra K. Verma et al.[31] proposed an animal detection technique by using a CNN model known as ResNet. They have utilized camera trap dataset, which contains two categories, namely, positive images, which include animals, and the negative images, which does not have the wild animals. They have utilized 90% data for the purpose of training and 10% data for testing purpose and get 91.4% accuracy. The brief review for animal detection is illustrated in Table 2.2.

Table 2.2 Summary of existing surveys related to wild animal dataset.

No.	Author	Year	Methods	Description
1	K. Figueroa et al.[24]	2014	SURF	Detecting animals in camera trap images
2	M. Pietrasik et al.[25]	2020	RetinaNet and YOLO v3	Animal detection
3	J. Parham et al.[26]	2018	DCNN and YOLO	Animal recognition pipeline
4	A. Loos et al.[27]	2018	YOLO v2 and SSD	Detect and classify different animal species
5	W.H.S. Antonio et al.[29]	2019	K-nearest neighbors and random forest	Warn the drivers on roads to detect the presence of the animals
6	B. Kellenberger et al.[30]	2019	Transfer sampling	Detect wildlife animals in unmanned aerial vehicle
7	G.K. Verma et al.[31]	2018	ResNet	Detecting wildlife animals in highly cluttered images

4.3.1.1 Experimental details

In this chapter, the images are obtained from the camera trap dataset[32] and captured in both infrared and visual formats, and the background in these images is highly cluttered. This dataset contains approximately 1850 images, and the dimension of each image varies in its size. The size of the image varies from 1920×1080 to 2048×1536, but they are finally resized to 224×224. They are categorized into two classes, i.e., positive images and negative images. The positive images contain the wild animals with different background, and in the negative images, the wild animals are not present. From this dataset, 75% of images are captured for training purpose, and the rest 25% are utilized for the testing purpose. The overall number of epochs taken is 250, and the learning rate is 10^{-3}. Fig. 2.3 shows the two classes of our dataset, containing both positive images and negative images.

4.3.1.2 Result analysis for animal dataset

Finally, we represent the qualitative results on the wild animal dataset with DenseNet model. Fig. 2.4 shows training results obtained from two different classes of the given dataset.

(a)

(b)

Figure 2.3 Sample Images of Wild animal Dataset: (A) Positive and (B) Negative. *From Zhang Z, He Z, Cao G, Cao W. Animal detection from highly cluttered natural scenes using spatiotemporal object region proposals and patch verification.* IEEE Trans Multimedia. *2016;18(10):2079–2092. https://doi.org/10.1109/TMM.2016.2594138.*

The accuracy of our proposed framework is described as the predicted image versus the actual image. Accuracy obtained on this datasets is 95.022%. Fig. 2.5 displays the results procured from the testing phase which is described in terms of loss and probability ratio.

Finally graphs are drawn to show the performance evaluation, which are shown in Fig. 2.6, Fig. 2.7, and Fig. 2.8.

Figure 2.4 Training result for animal dataset. *From Zhang Z, He Z, Cao G, Cao W. Animal detection from highly cluttered natural scenes using spatiotemporal object region proposals and patch verification.* IEEE Trans Multimedia. *2016;18(10):2079–2092. https://doi.org/10.1109/TMM.2016.2594138.*

First graph is drawn between loss versus epoch and epochs, which present the accuracy of the dataset for the given epochs. Next graph is drawn between error versus epochs and epochs, which represent the percentage of error in the given epochs. The final graph is drawn between loss versus epoch and epochs, which present the percentage of loss on the given epochs.

4.3.2 Knife dataset

Michał Grega et al.[33] proposed automatic detection method for firearms and knives captured from CCTV systems. They have used a three-layer neural network and support vector machine[34] parameters for the detection process. By utilizing the edge histogram features, they have attained an accuracy of

Figure 2.5 Result on Animal dataset in terms of prediction/actual/loss/probability. *From Zhang Z, He Z, Cao G, Cao W. Animal detection from highly cluttered natural scenes using spatiotemporal object region proposals and patch verification.* IEEE Trans Multimedia. *2016;18(10):2079–2092. https://doi.org/10.1109/TMM.2016.2594138.*

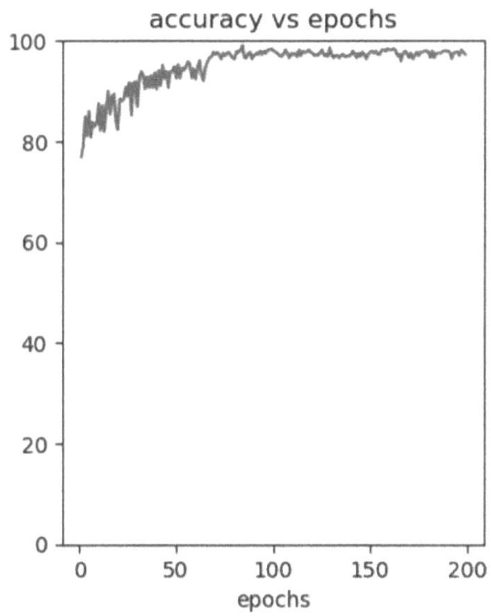

Figure 2.6 Representation of the performance evaluation for animal dataset: accuracy versus epochs.

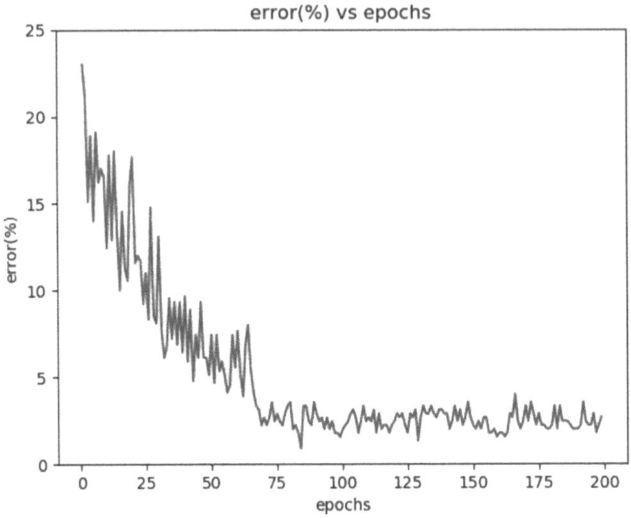

Figure 2.7 Representation of the performance evaluation for animal dataset: error versus epochs.

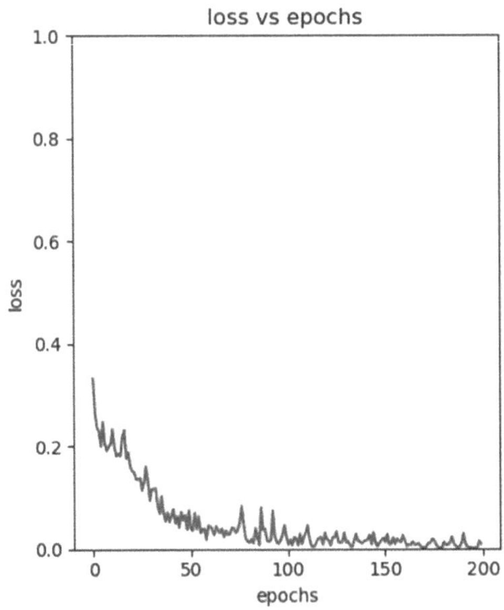

Figure 2.8 Representation of the performance evaluation for animal dataset: loss versus epochs.

Table 2.3 Summary of existing surveys related to knife dataset.

No.	Authors	Year	Methods	Description
1	M. Grega et al.[33]	2016	Fuzzy classification and support vector machine	Detecting dangerous situations in CCTV
2	A. Glowacz et al.[35]	2015	Active appearance models	Knife detection
3	R. Guo et al.[36]	2019	SSD-ResNet101	Detection of small tools in X-ray security inspection
4	A. Warsi et al.[38]	2020	Faster R-CNN and YOLO	Detecting handguns and knives in ImageNet dataset

91% and only 5% of the false-positive rate. Andrzej Glowacz et al.[35] propounded a knife detection framework by utilizing the active appearance models. Ruihong Guo et al.[36] propounded a knife detection and identification methodology. In this chapter, an improved version of the SSD model has been utilized. VGG16[37] of SSD with ResNet101 model is replaced by SSD ResNet101 model and attained a mAP of 89.7% on the Specialized Dataset for Control Knives 2019 (SDCK 2019). Arif Warsi et al.[38] propounded a detection technique for handgun and knife by utilizing Faster R-CNN and YOLO. They have used the ImageNet dataset for comparing their results and got an accuracy of 93% by using Faster R-CNN. The summary of existing surveys on knife detection is illustrated in Table 2.3.

4.3.2.1 Experimental details

The complete database is publicly available at Ref. 39 and it contains the images of knife, held by the human hand. This dataset contains two categories of images, namely, positive and negative images, as shown in Fig. 2.9.

The images were taken in the indoor area, so some images contain nuisance in the form of reflection and type of knife. In our chapter, we have used a total number of 1560 images, out of which 1010 images are taken as positive and 550 images are taken as negative. Size of the images is taken as 224 × 224, and 75% images are utilized for training and 25% for the testing.

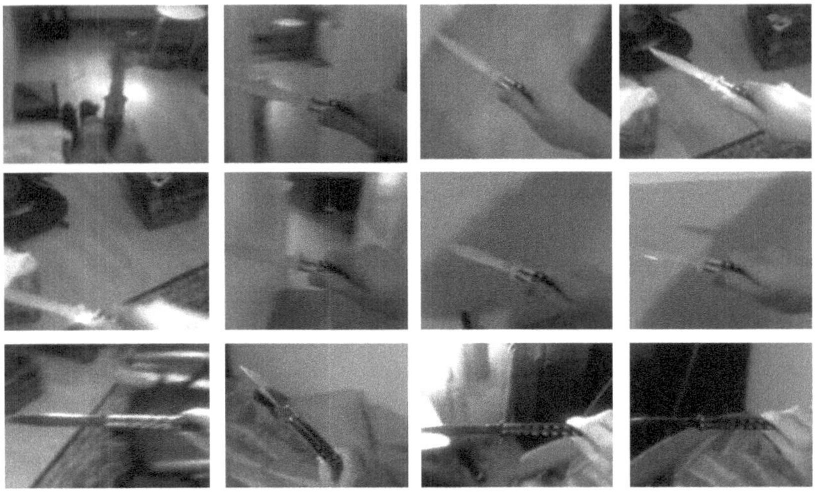

(a)

(b)

Figure 2.9 Sample images of knife dataset: (A) positive and (B) negative. *From* Knife Image Database. *2015. https://dx.doi.org/10.1007/s11042-015-2697-z.*

4.3.2.2 Result analysis for knife dataset

The qualitative results for knife dataset with DenseNet model are described here. Fig. 2.10 shows training results obtained from two different classes of the given dataset.

The accuracy of the propounded framework is described in terms of the predicted image versus the actual image. Accuracy obtained for the given dataset is 95.209% with the given DenseNet-161 model. Fig. 2.11 shows the results obtained from the testing phase and shows the loss and the probability ratio.

The graphs for the performance evaluation of knife dataset are drawn in Fig. 2.12, Fig. 2.13, and Fig. 2.14.

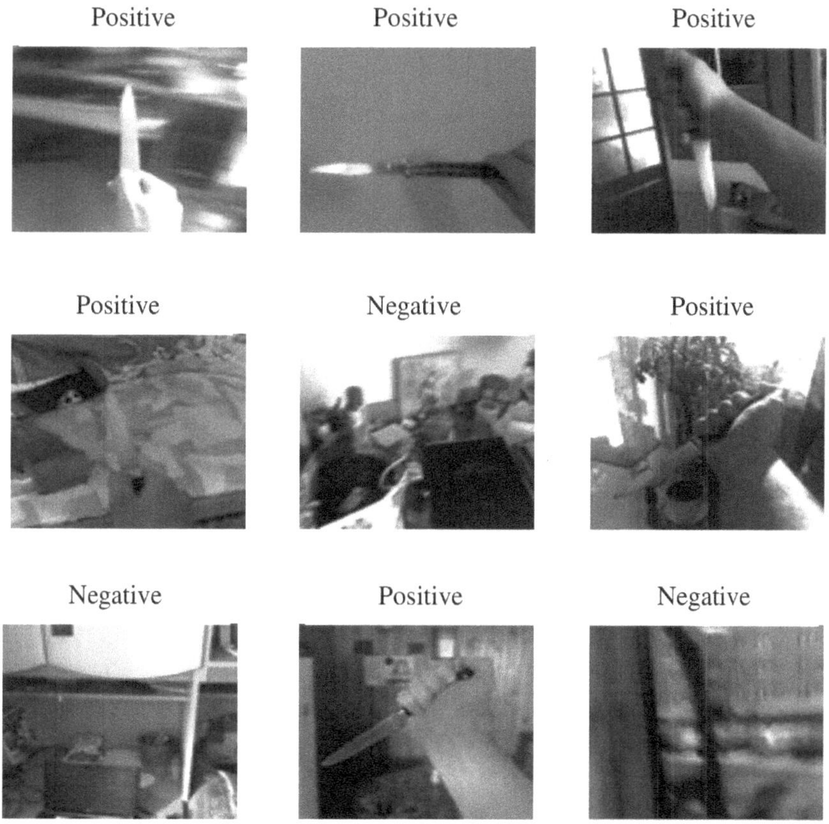

Figure 2.10 Training result for knife dataset. *From* Knife Image Database. *2015. https://dx.doi.org/10.1007/s11042-015-2697-z.*

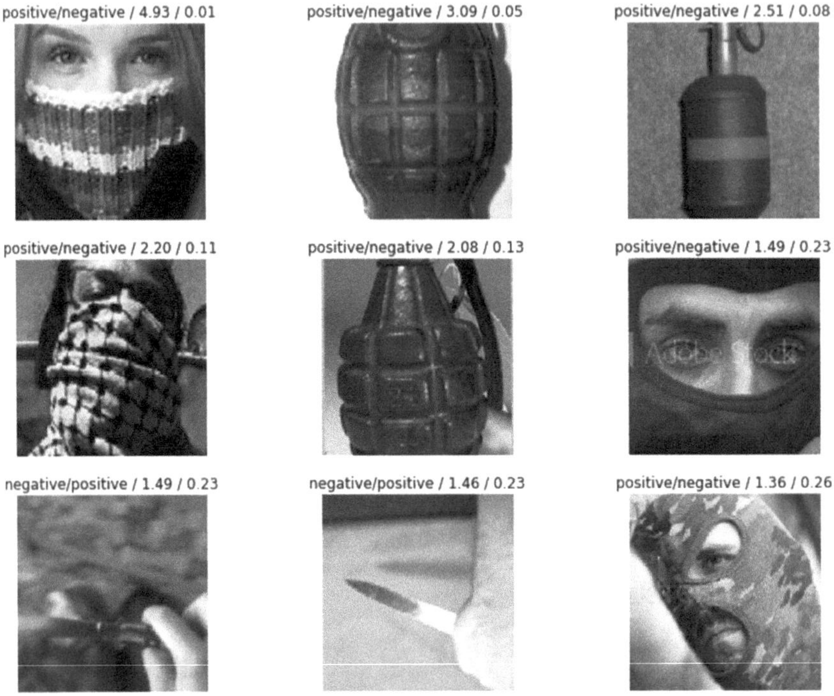

Figure 2.11 Result on knife dataset in terms of prediction/actual/loss/probability. *From* Knife Image Database. *(2015). https://dx.doi.org/10.1007/s11042-015-2697-z.*

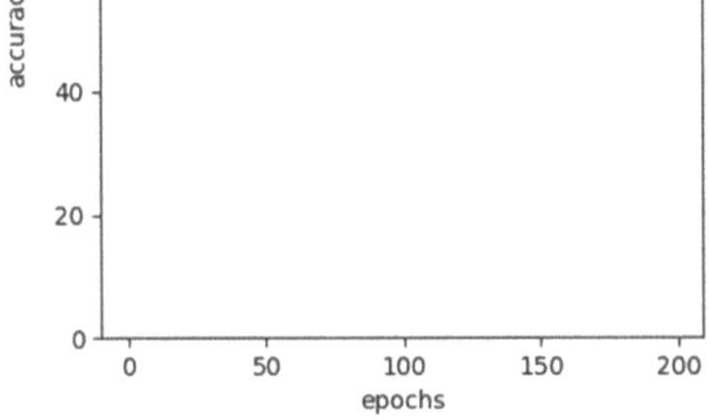

Figure 2.12 Representation of the performance evaluation for knife dataset: accuracy versus epochs.

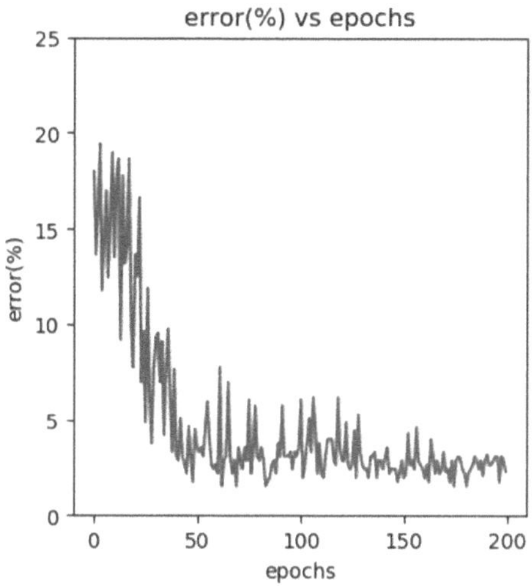

Figure 2.13 Representation of the performance evaluation for knife dataset: error versus epochs.

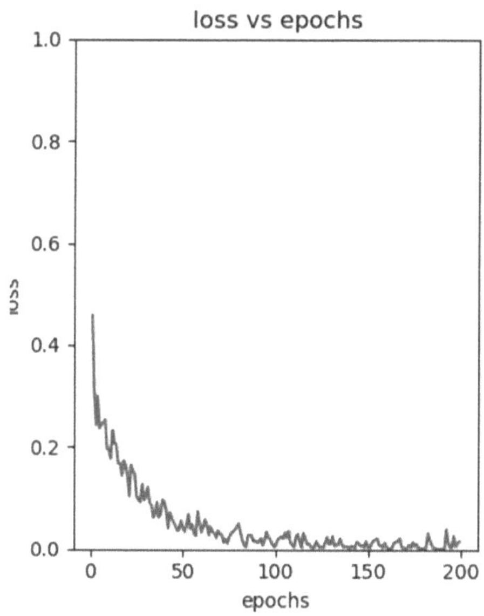

Figure 2.14 Representation of the performance evaluation for knife dataset: loss versus epochs.

5. Result analysis and discussion

The DenseNet-161 model was trained and tested on two databases, namely wild animal and knife. The result are presented in Table 2.4: in the first column, the datasets are described, and in the second one, the total number of classes in both the datasets is depicted. These datasets contain two classes, i.e., positive class and negative class. Positive class includes the particular object present in that image, whereas that object is missing in the negative class. The third column contains the total number of images, and in the fourth column total time (in minutes) taken by the datasets during the training time is elucidated. The last column exhibits the percentage of accuracy achieved by these datasets.

The confusion matrix[40] is the one among the foremost and widely used performance measuring techniques for classification models. It is also called as an error matrix, where the rows indicate the ground truth values and columns indicate the classification done by classifiers as shown in Fig. 2.15 and Fig. 2.16.

For the classification issue of n-class, the confusion matrix is an $n \times n$ matrix. With this matrix, the correct recognition rate R_i and the error recognition rate W_i of the data objects can be calculated.

$$R_i = \frac{n_i}{\sum_{j1}^{k} n_{ij}}$$

Here the category set is defined as $C = C_1, C_2 \ldots C_k$, and among them, n_{ii} is the precisely recognized numbers of confusion matrix and n_{ij} denotes that C_i is recognized as C_j by classifier.

5.1 Performance comparison

The quantitative analysis shows that our proposed framework attains an excellent advancement in the detection of given datasets. DenseNet model is identified as an advantageous framework for those vision tasks which are easily perceptible. This section presents the comparison of our proposed results using the DenseNet method with the other techniques, including Faster R-CNN,[41] ResNet,[42] BCNet,[43] and CNN models.[44] The accuracy rate comparison of various studies along with the proposed method is given in Table 2.5. Bin Jhiang et al. reported 85.6% accuracy on animal dataset with the BCNet, whereas Mustafa Kaya et al. have achieved the accuracy of 94.37% on the criminal tools like gun, knife, etc., by utilizing the CNN model. Table 2.5 clearly implies that our method is outperforming most of the methods in the class, by accurately identifying the correct region of interest.

Table 2.4 Detection performance of DenseNet with different datasets.

Dataset	Number of classes		Total images	Time taken (in minutes)	Accuracy (%)
	Negative images	Positive images			
Wild animal	508	1298	1806	1:51:26	95.02
Knife	550	1010	1560	2:55:13	95.20

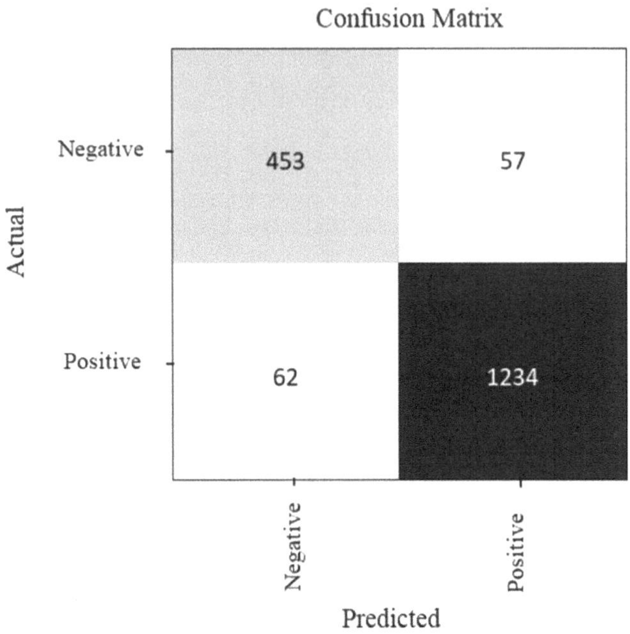

Figure 2.15 Confusion matrix for animal dataset.

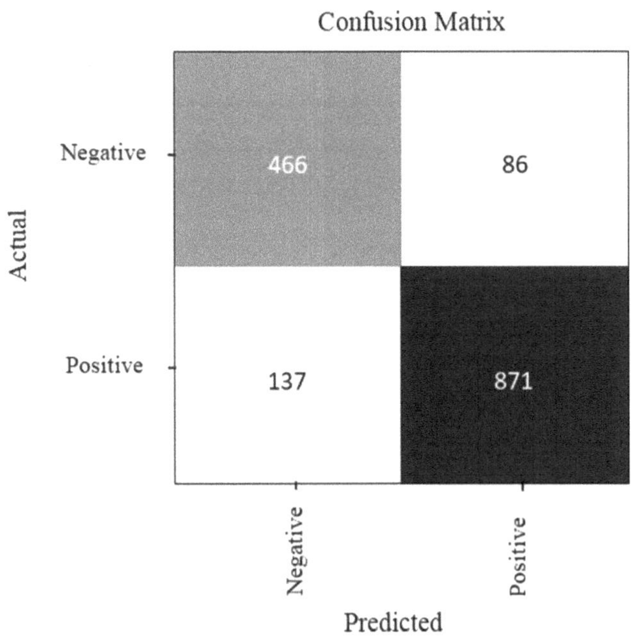

Figure 2.16 Confusion matrix for knife dataset.

Table 2.5 Comparison of given datasets with other methods.

Author	Approach	Classification accuracy
Our method (animal dataset)	DenseNet-161	95.02%
Our method (Knife dataset)	DenseNet-161	95.20%
Verma and Dhillon[41]	Faster R-CNN	92.6%
Verma and Gupta[42]	ResNet	91.2%
B. Jhiang et al.[43]	BCNet	85.6%
M. Kaya et al.[44]	CNN	94.37%

6. Conclusion

In this work, we proposed an object classification and recognition framework using DenseNet-161 model. The experiments are performed on two datasets, namely, wild animal camera trap and knife. Our basic idea is to extract the features by categorizing two classes of datasets, i.e., positive and negative. In positive class, the object is present, whereas in the negative class, no object is present in the given image. After that we extract the features and classify the objects using DenseNet-161 model. We claim two major contributions. First, a systematic framework is established to recognize and classify objects in camera trap animal and knife datasets. Second, we extricate the features and classify images utilizing DenseNet-161 model. With the proposed methodology, we have obtained an accuracy of 95.02% and 95.20% on camera trap wild animal and knife dataset, respectively. Experimental results show that when compared with other methodologies, our method attains maximum accuracy. In the future, we aim to examine large-scale and challenging datasets for image classification and detection problems. Also, we will concentrate on optimizing a fusion of CNN pretrained models to further improve the robustness of object classification.

References

1. Khan A, Sohail A, Zahoora U, Qureshi AS. A survey of the recent architectures of deep convolutional neural networks. *Artif Intell Rev*. 2020;53(8):5455–5516. https://doi.org/10.1007/s10462-020-09825-6.
2. Liu L, Ouyang W, Wang X, et al. Deep learning for generic object detection: a survey. *Int J Comput Vis*. 2020;128(2):261–318. https://doi.org/10.1007/s11263-019-01247-4.
3. Tong K, Wu Y, Zhou F. Recent advances in small object detection based on deep learning: a review. *Image Vis Comput*. 2020;97. https://doi.org/10.1016/j.imavis.2020.103910.

4. Wang Y, Li H, Jia P, Zhang G, Wang T, Hao X. Multi-scale densenets based aircraft detection from remote sensing images. *Sensors*. 2019;19(23).
5. Cruciani F, Vafeiadis A, Nugent C, et al. Feature learning for human activity recognition using convolutional neural networks. *CCF Trans Pervasive Comput Interact*. 2020; 2(1):18–32. https://doi.org/10.1007/s42486-020-00026-2.
6. Roy SK, Krishna G, Dubey SR, Chaudhuri BB. HybridSN: exploring 3-D-2-D CNN feature hierarchy for hyperspectral image classification. *Geosci Rem Sens Lett IEEE*. 2020; 17(2):277–281. https://doi.org/10.1109/LGRS.2019.2918719.
7. Zhao ZQ, Zheng P, Xu ST, Wu X. Object detection with deep learning: a review. *IEEE Transact Neural Networks Learn Syst*. 2019;30(11):3212–3232. https://doi.org/10.1109/TNNLS.2018.2876865.
8. Huang G, Liu Z, Van Der Maaten L, Weinberger KQ. Densely connected convolutional networks. In: *Proceedings - 30th IEEE Conference on Computer Vision and Pattern Recognition, CVPR 2017*. Vol. 2017. 2017:2261–2269. https://doi.org/10.1109/CVPR.2017.243.
9. Krizhevsky A, Sutskever I, Hinton GE. ImageNet classification with deep convolutional neural networks. *Adv Neural Inf Process Syst*. 2012;2:1097–1105.
10. Zhang K, Guo Y, Wang X, Yuan J, Ding Q. Multiple feature reweight DenseNet for image classification. *IEEE Access*. 2019;7:9872–9880. https://doi.org/10.1109/ACCESS.2018.2890127.
11. Zhang K, Guo Y, Wang X, Yuan J, Ma Z, Zhao Z. Channel-wise and feature-points reweights Densenet for image classification. In: *Proceedings - International Conference on Image Processing, ICIP*. 2019. 2019:410–414. https://doi.org/10.1109/ICIP.2019.8802982.
12. Hu J, Shen L, Sun G. Squeeze-and-excitation networks. *IEEE Comput Soc Conf Comput Vis Pattern Recogn*. 2018:7132–7141. https://doi.org/10.1109/CVPR.2018.00745.
13. Zhai S, Shang D, Wang S, Dong S. DF-SSD: an improved SSD object detection algorithm based on DenseNet and feature fusion. *IEEE Access*. 2020;8:24344–24357. https://doi.org/10.1109/ACCESS.2020.2971026.
14. Liu W, Anguelov D, Erhan D, et al. SSD: Single shot multibox detector. *Lect Notes Comput Sci*. 2016;9905:21–37. https://doi.org/10.1007/978-3-319-46448-0_2.
15. Hao W, Zhang Z. Spatiotemporal distilled dense-connectivity network for video action recognition. *Pattern Recogn*. 2019;92:13–24. https://doi.org/10.1016/j.patcog.2019.03.005.
16. Kreso I, Krapac J, Segvic S. Efficient ladder-style DenseNets for semantic segmentation of large images. *IEEE Trans Intell Transport Syst*. 2021;22(8):4951–4961. https://doi.org/10.1109/tits.2020.2984894.
17. He K, Zhang X, Ren S, Sun J. Spatial pyramid pooling in deep convolutional networks for visual recognition. *IEEE Trans Pattern Anal Mach Intell*. 2015;37(9):1904–1916. https://doi.org/10.1109/TPAMI.2015.2389824.
18. T T, G L, X L, Q G. Image super-resolution using dense skip connections. In: *Proceedings of the IEEE International Conference on Computer Vision*. 2017:4799–4807.
19. Wang W, Li Y, Zou T, Wang X, You J, Luo Y. A novel image classification approach via dense-mobilenet models. *Mobile Inf Syst*. 2020:2020. https://doi.org/10.1155/2020/7602384.
20. Howard A, Zhu M, Chen B, et al. *Efficient Convolutional Neural Networks For Mobile Vision Applications*. 2017.
21. Shamsolmoali P, zareapoor M, Zhang J, Yang J. Image super resolution by dilated dense progressive network. *Image Vis Comput*. 2019;88:9–18. https://doi.org/10.1016/j.imavis.2019.03.006.

22. Huang G, Liu S, Maaten LVD, Weinberger KQ. CondenseNet: an efficient DenseNet using learned group convolutions. *IEEE Comput Soc Conf Comput Vis Pattern Recogn.* 2018:2752–2761. https://doi.org/10.1109/CVPR.2018.00291.
23. Gupta P, Verma GK. Wild animal detection using discriminative feature-oriented dictionary learning. In: *Proceeding - IEEE International Conference on Computing, Communication and Automation, ICCCA 2017.* Vol. 2017. 2017:104–109. https://doi.org/10.1109/CCAA.2017.8229781.
24. Figueroa K, Camarena-Ibarrola A, García J, Villela HT. Fast automatic detection of wildlife in images from trap cameras. *Lect Notes Comput Sci.* 2014;8827:940–947. https://doi.org/10.1007/978-3-319-12568-8_114.
25. Singh A, Pietrasik M, Natha G, Ghouaiel N, Brizel K, Ray. Animal Detection in man-made environments. In: *Proceedings of the IEEE/CVF Winter Conference on Applications of Computer Vision.* 2020:1438–1449.
26. Parham J, Stewart C, Crall J, Rubenstein D, Holmberg J, Berger-Wolf T. An animal detection pipeline for identification. In: *Proceedings - 2018 IEEE Winter Conference on Applications of Computer Vision, WACV 2018.* 2018:1075–1083. https://doi.org/10.1109/WACV.2018.00123.
27. Loos A, Weigel C, Koehler M. Towards automatic detection of animals in camera-trap images. In: *European Signal Processing Conference.* 2018. 2018:1805–1809. https://doi.org/10.23919/EUSIPCO.2018.8553439.
28. Redmon J, Divvala S, Girshick R, Farhadi A. You only look once: unified, real-time object detection. *IEEE Comput Soc Conf Comput Vis Pattern Recogn.* 2016;2016: 779–788. https://doi.org/10.1109/CVPR.2016.91.
29. Antônio WHS, Da Silva M, Miani RS, Souza JR. A proposal of an animal detection system using machine learning. *Appl Artif Intell.* 2019;33(13):1093–1106. https://doi.org/10.1080/08839514.2019.1673993.
30. Kellenberger B, Marcos D, Lobry S, Tuia D. Half a percent of labels is enough: efficient animal detection in UAV imagery using deep CNNs and active learning. *IEEE Trans Geosci Rem Sens.* 2019;57(12):9524–9533. https://doi.org/10.1109/TGRS.2019.2927393.
31. Dhillon A, Verma GK. Wild animal detection from highly cluttered forest images using deep residual networks. *Lect Notes Comput Sci.* 2018;11278:230–238. https://doi.org/10.1007/978-3-030-04021-5_21.
32. Zhang Z, He Z, Cao G, Cao W. Animal detection from highly cluttered natural scenes using spatiotemporal object region proposals and patch verification. *IEEE Trans Multimed.* 2016;18(10):2079–2092. https://doi.org/10.1109/TMM.2016.2594138.
33. Grega M, Matiolański A, Guzik P, Leszczuk M. Automated detection of Firearms and knives in a CCTV image. *Sensors.* 2016;16(1):47. https://doi.org/10.3390/s16010047.
34. Evgeniou T, Pontil M. Support vector machines: theory and applications. In: *Advanced Course on Artificial Intelligence.* 1999:249–257.
35. Glowacz A, Kmieć M, Dziech A. Visual detection of knives in security applications using active appearance models. *Multimed Tool Appl.* 2015;74(12):4253–4267. https://doi.org/10.1007/s11042-013-1537-2.
36. Guo R, Zhang L, Ying Y, Sun H, Han Y, Tan H. Automatic detection and identification of controlled knives based on improved SSD model. In: *Proceedings - 2019 Chinese Automation Congress, CAC 2019.* 2019:5120–5125. https://doi.org/10.1109/CAC48633.2019.8996879.
37. Simonyan K, Zisserman A. *Very Deep Convolutional Networks for Largescale Image Recognition.* 2014.
38. Warsi A, Abdullah M, Husen MN, Yahya M. Automatic Handgun and knife detection algorithms: a review. In: *Proceedings of the 2020 14th International Conference on Ubiquitous Information Management and Communication, IMCOM 2020.* 2020. https://doi.org/10.1109/IMCOM48794.2020.9001725.

39. *Knife Image Database*. 2015. https://doi.org/10.1007/s11042-015-2697-z.
40. Xiong Y. Building text hierarchical structure by using confusion matrix. In: *2012 5th International Conference on Biomedical Engineering and Informatics, BMEI 2012*. 2012: 1250–1254. https://doi.org/10.1109/BMEI.2012.6513202.
41. Verma GK, Dhillon A. A handheld gun detection using faster R-CNN deep learning. In: *ACM International Conference Proceeding Series*. 2017:84–88. https://doi.org/10.1145/3154979.3154988.
42. Verma VG, Gupta P. Wild animal detection from highly cluttered images using deep convolutional neural network. *Int J Comput Intell Appl*. 2018;17(04):1850021.
43. Jiang B, Huang W, Tu W, Yang C. An animal classification based on light convolutional network neural network. In: *Proceedings - 2019 International Conference on Intelligent Computing and its Emerging Applications, ICEA 2019*. 2019:45–50. https://doi.org/10.1109/ICEA.2019.8858309.
44. Kaya M, Karakuş BA, Karakuş S. Binary classification of criminal tools from the images of the case using CNN. In: *2018 International Conference on Artificial Intelligence and Data Processing (IDAP)*. 2018:1–6.

CHAPTER THREE

Deep learning-based image processing for analyzing combustion behavior of gel fuel droplets

Vandan Agarwal[1], Puneet Chitkariya[1], Ankur Miglani[1], Purushothaman Nandagopalan[2], Jerin John[3], Pavan Kumar Kankar[4]

[1]Microfluidics and Droplet Dynamics Lab, Department of Mechanical Engineering, Indian Institute of Technology, Indore, Madhya Pradesh, India
[2]School of Aeronautical Sciences, Hindustan Institute of Technology and Science, Chennai, Tamil Nadu, India
[3]Perigee Aerospace Inc, Daejeon, Republic of Korea
[4]System Dynamics Lab, Department of Mechanical Engineering, Indian Institute of Technology, Indore, Madhya Pradesh, India

Contents

Smart Electrical and Mechanical Systems
ISBN: 978-0-323-90789-7
https://doi.org/10.1016/B978-0-323-90789-7.00011-7

1. Introduction

The feasibility of next-generation rocket propulsion, including sounding rockets and tactical missiles, will depend on their adaptability to environmentally friendly fuels that offer high energy density and low ignition delay, while being easy to store and handle.[1–3] In this regard, the high-performance gelled fuels have emerged as viable alternatives to conventionally used neat propellants. This is due to their shear-thinning behavior that combines the desired properties of both solid and liquid propellants that are required in practice. For instance, gel fuels allow easy storability (low or negligible shear condition) like solid fuels, thereby reducing the chances of a hazard due to an accidental spill or leakage. Alternatively, liquid-like behavior under high-shear in injectors would allow thrust modulation and reignitability.[4–7]

Despite the well-known advantages of gel fuels, their combustion behavior is poorly understood, even at the isolated droplet scale, which represents the subgrid unit of a spray. This is due to the multicomponent and multiscale nature of the thermo-physical mechanisms, and their complex interplay that govern the combustion behavior of gel fuels. Combustion of gel fuel droplets is highlighted by disruptive vapor jetting events that occur asymmetrically around the droplet and randomly during the droplet lifetime. These events occur at extremely short timescales ~ O (1–10 ms), and across length scales ranging from ~ O (100 μm) for events occurring at the subdroplet scale (i.e., droplet interior) up to the flame scale ~2 mm.[8] Furthermore, these events lead to bulk droplet shape oscillations featuring aperiodic swell-shrink cycles, and at times sudden bursting of the droplet, which results in severe shape deformation. To determine the burn rates of gel fuel droplets that exhibit such complex combustion behavior, most studies rely on the conventional MATLAB or Image-J based postprocessing of the back-lit droplet images, where a thresholding-based edge-detection is used to identify droplet boundary. Subsequently, the area enclosed by this droplet boundary is used to determine the projected area-based equivalent diameter for each frame, and hence the variation of droplet diameter with time.[9–11] However, to capture the rapidly changing droplet shape due to jetting, the images are often captured at high frame rates ~ O (10,000 fps) using a monochrome ultrahigh-speed cameras, thereby limiting the image resolution to less than one megapixel. In addition to the reduced image quality, fluctuations in light intensity occurring at the electrical frequency (50–60 Hz; depending on the source of power supply) are also captured by the camera sensor, which leads to a variation in the image brightness from one frame to another. Furthermore, the physical size of the droplet (often less than 1 mm) demands high-magnification images with an optical resolution ~ O (5 μm/pixel), which in turn restricts the depth-of-field.[12,13] A shallow depth-of-field combined with a low resolution along with a dynamically changing droplet shape (in 3D) results in a compromise with the sharpness, contrast, and intensity fluctuations, which makes the droplet edge at times indistinguishable from the background. This makes the task of accurately detecting the droplet edge and its boundary inherently challenging.

Conventional edge detection techniques, popularly known as the structured-edge detection, determine the gradient magnitude of the pixel values to identify the local maxima, which is a measure of the edge strength.[14] One such method, the Canny edge detection[15] uses a filter based on the derivative of a Gaussian to compute the intensity of the gradients. However, this method relies on visual validation and manual feedback of the threshold values for the intensity gradient to retain the edges. Therefore, structured-edge detection methods require manual intervention, which would be time-intensive, and therefore, ineffective for large datasets. This drawback leaves these methods unfit for high-speed combustions studies that generate tens of thousands of images in one experimental trial.

This chapter describes an algorithm that overcomes the above difficulties in specifically demarcating the droplet boundary and extracting geometric parameters from it. The algorithm is dissolved into several stages, which are sequentially applied to every single frame of the acquired combustion recording. Detecting edges in a raw image is the first step to locate the droplet. Conventional edge detection methods like canny edge detection struggle to solve the problem due to the poor sharpness and high brightness fluctuation between frames. To alleviate this challenge, a deep learning based holistically nested edge detection (HED)[16] method is used to generate nonbinary edge maps corresponding to each frame of the video. The nonbinary edge maps produced by the HED are discontinuous and rich in turbulence and therefore require restructuring. For noise removal and droplet localization, binary bit-masking and color thresholding at a predetermined threshold was applied. The resulting image gives a set of points surrounding the droplet. These points are then filtered, based on their location, into boundary or nonboundary points, the latter being discarded. The boundary points are now used to create a binary map, which is then used to calculate the number of related components using contour detection techniques. In the case of nonsingular components, a droplet envelope is fabricated by performing a convex hull operation[17] on the set of boundary points. Finally, a series of morphological transformations are equipped to achieve a refined and smooth droplet boundary.

2. Experimental methodology

2.1 Fuel formulation

In this study, the combustion behavior of nonmetallized gel fuel droplets is investigated, which comprise of three components: (1) Research grade ethanol (99.8% Pure; CAS No. 64-17- 5) which forms the base fuel, (2) macromolecular Hydroxypropyl Methyl Cellulose (HPMC: CAS No. 9004-65-3 and density 689.18 kg/m^3) or Methyl Cellulose (MC: CAS No. 9004-67-5 and density 503.92 kg/m^3), which are chosen as organic gellants (OG) or gelling agents, and (3) deionized water, which forms the base solvent for the gellant. A stable gel fuel is prepared through a three-step process as described in Ref. 2: First, the gelling agent is added to ethanol and hand stirred vigorously at room temperature for approximately 120 s. Second, the deionized water is added to the ethanol-gellant mixture, which is then stirred vigorously using an impeller-based mechanical mixer at 500 rpm for ~180 s. Finally, the resulting mixture is left undisturbed at room temperature to initiate the process of gel network formation. A stable gel fuel is formed completely is approximately 48 h. The composition of these gel fuels fuel formulations having different gellant concentrations is detailed in Table 3.1.

Table 3.1 Composition of organic gellant-laden gel fuels.

Gel fuel	Ethanol (wt.%)	HPMC (wt.%)	MC (wt.%)	Water (wt.%)
HPMC 3	82	3	–	15
HPMC 4	81	4	–	15
HPMC 5	85	5	–	10
HPMC 6	84	6		10
MC 8	76	–	8	16
MC 9	77	–	9	14

2.2 Test facility

The experimental facility for capturing the burning behavior of gel fuel droplets using high speed imaging.[18,19] A volume calibrated μ-syringe with a 30-gauge needle is used to dispense a gel fuel droplet of ~3 μL at the end of an 80 μm fused quartz fiber. Since the diameter of the suspension fiber is less than 100 μm, and fused quartz has a low thermal conductivity of ~1.4 W/m-K at 298 K, the physical or thermal interference due to the fiber can be neglected. Droplet ignition is achieved via a Nichrome wire heater that is brought close to the droplet using a solenoid. The solenoid is programmed to retract the wire heater after a short duration of 30 ms. The heat input to the wire required to ignite the droplet is regulated by a DC electric power source. The ignition and the combustion of ethanol-gel droplets in the pendant mode are captured using an ultrahigh-speed camera at ambient temperature and pressure conditions.

2.3 Image acquisition

The dynamics of a burning Ethanol-Water-HPMC/MC droplet is captured at 10,000 fps, with an exposure set at 100 μs, and at a spatial resolution of 4 μm/pixel. This is achieved using an ultrahigh-speed camera (Phantom VEO710L) with an attached 6.5 X zoom lens (Navitar 1−6265). Subsequently, the planar (2D) images of the droplet are postprocessed using a deep-learning approach to determine six geometrical parameters (see Fig. 3.1) for each frame, namely, droplet area, diameter, circumference, centroid, aspect ratio (maximum height/maximum width), and the distance from the droplet centroid to the droplet edge along any radial direction. The parameters are extracted for five combustion videos of six different gel fuels, namely, HPMC 3, HPMC 4, HPMC 5, HPMC 6, MC 8, and MC 9.

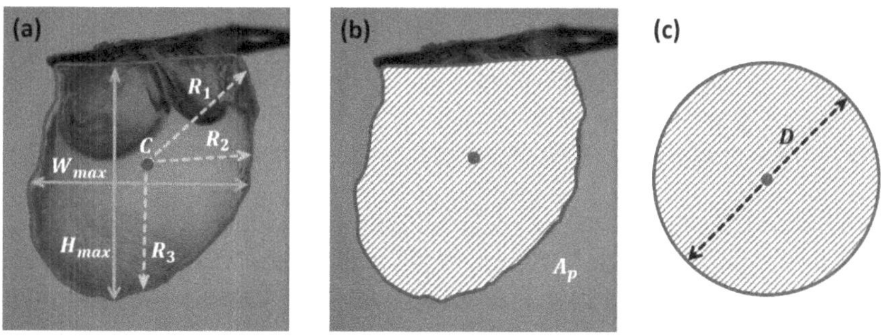

Figure 3.1 A schematic representation of the geometrical parameters of a gel fuel droplet. A schematic representation of the geometrical parameters required to characterize the disruptive combustion behavior (resulting from jetting of unreacted fuel vapors) of a gel fuel droplet in terms of its gasification or burn rate.

3. Edge detection methods

Detecting the droplet edges precisely in the raw images is the key step in reconstructing the 2D projection of the droplet shape and subsequently determining the geometrical parameters. This step simplifies the subsequent image processing by reducing the amount of data to be analyzed while retaining the requisite structural information about the droplet boundaries. However, edge detection is a challenging task due to several constraints, as discussed in Section 1, including the low image resolution, a narrow depth-of-field at high magnifications, variation in the brightness from one frame to another, and the dynamically changing droplet shape.

3.1 Conventional methods

The conventional edge detection techniques used for capturing the droplet boundary determine the gradient magnitude of the pixel values to identify the local maxima by convolving various filters through the image matrix, which provides a measure of the edge strength. To test the effectiveness of conventional edge detection methods, the Canny edge detector is first applied to the burning droplet images, which is one of the most extensively used methods. By using the edge detection operators such as the Sobel,[20] this method computes the edge intensity in different directions from the gradient calculation. Subsequently, the magnitude and direction of edge strength for each of the pixels is estimated from these edge intensities. The detected boundaries are then thinned using the nonmaximum suppression, which preserves only those pixel values that have the largest magnitude in the gradient direction among its surrounding pixels. By using double thresholds whose values are provided manually, these pixel values are classified into three categories: strong, weak, and nonrelevant. The nonrelevant pixels are simply eliminated, while the weak pixels are retained only if they are surrounded by one or more strong pixels in the final output.

The Canny edge detector relies on experimentation, visual validation, and manual feedback of threshold values to determine the droplet edges. As discussed in Section 1, since the images captured at high-speed incur fluctuations in contrast and brightness across the frames, a single value of threshold cannot be considered for an entire video. Furthermore, the Canny edge detection relies on several preprocessing steps such as image conversion to grayscale or binary format, and subsequent image enhancement through smoothening and blurring to generate an accurate edge map. Fig. 3.2 shows

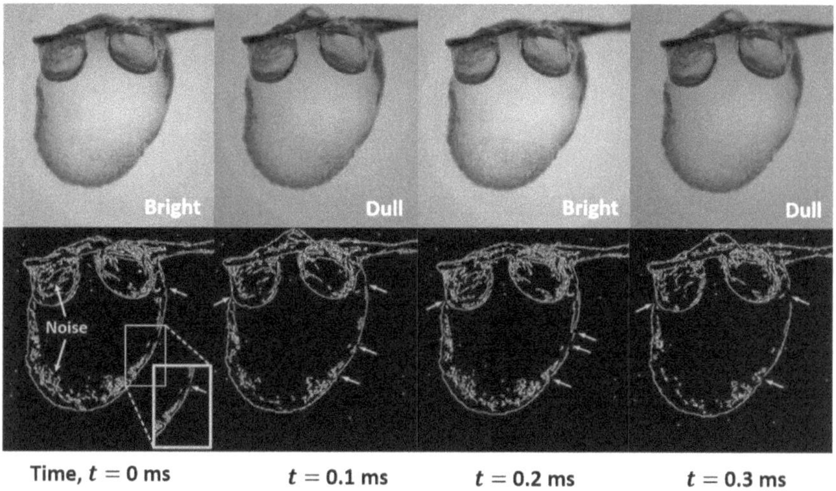

Figure 3.2 Four consecutive images obtained from the gel droplet combustion recordings. Four consecutive images obtained from the gel droplet combustion recordings: (Top row) raw images (Bottom row) corresponding processed images using Canny Edge Detection.

the processed images for four consecutive frames in the droplet combustion recording that are obtained by applying the Canny edge detection technique to the raw images. It is apparent that the edge maps generated by Canny edge detectors are binary, which makes it difficult to differentiate the droplet boundary from the background noise. Furthermore, a single threshold value cannot accommodate a large variation in the intensity across frames, which results in edge discontinuity at multiple locations. Discontinuity in the edge occurs at several places along the droplet boundary, as shown marked by an arrow in Fig. 3.2. Also, the processed images contain a lot of noise near the droplet edge due to the pattern and texture of the droplet surface; formed by the nonuniform gellant layer. Since high-speed imaging generates thousands of frames, the limitations of Canny edge detection make their postprocessing time-consuming, and therefore, ineffective for large datasets. Furthermore, the Canny edge detection works based on predetermined filters, where threshold value needs to be tuned manually, and therefore, it does not understand the context of the image. In contrast, the deep learning-based detectors develop these filters based on the layer-by-layer learning of the features, and hence, offer an attractive alternative approach for edge detection.

3.2 Deep learning-based edge detection

Deep learning-based image processing that adapts hierarchical convolutional neural networks (CNN)-based training can address the challenges faced by conventional image processing techniques. The deep learning-based approach takes an RGB image as an input and assigns importance (learnable weights and biases) to different aspects and objects in the image, thus allowing them to be distinguished.[21–24] While the filters are required to be hand-engineered in conventional approaches, a CNN can learn these filters on its own with adequate training and produce an edge map as an output while preserving the droplet boundaries in the image. Furthermore, the amount of preprocessing required by a CNN is significantly less compared to the conventional methods. Therefore, CNN-based methods have shown promising improvements over conventional methods such as the Canny Edge Detectors.

The concept of multiscale and multilevel feature learning is critical to the understanding of hierarchical learning in a deep CNN. Here, a rectangular subregion of the input image, known as the receptive field of that layer, has an impact on the feature maps created by that layer. The feature representation in a CNN is multiscaled in the form of increasingly larger receptive fields spread across the layers. In other words, the initial convolutional layers of the network generate feature maps that encode the basic semantic information extracted from small regions of the image, whereas the deeper layers encode the more complex semantic information based on the representative features from larger portions of the image or the entire image itself. The edge detection accuracy of a CNN-based architecture, which is a binary classification model, is measured by computing the F-score. F-score is calculated from the harmonic mean of the precision and a recall of the model on a dataset.[25] The maximum value it can approach is unity, which represents a perfectly accurate model. Previously, the results produced by fully convolutional networks (FCN) were less satisfactory, for example, the F-score on BSDS500 dataset[26] is 0.745. However, an integration of deep supervision during network training has been the focus of recent CNN-based edge prediction algorithms. With the incorporation of this method, a more transparent and cohesive contribution is achieved from the multiscale responses of the hidden layers. While the recent CNN-based edge detection methods have shown promising improvement in performance compared to the conventional methods such as the Canny edge detection, structured edges, and sketch tokens, there is a vast scope of performance enhancement in terms of the F-score and the processing speed. Even with the current high-performance graphical processing units (GPUs), their image prediction speed can range up to a few hours. The holistically nested edge detection technique, which incorporates these characteristics, and shows a significant improvement in the F-score (0.782 on the BSDS500 dataset) is utilized for droplet edge detection in this study.

4. Holistically nested edge detection

HED has a unique ability to automatically learn the rich hierarchical features, which provides an end-to-end edge and boundary detection approach for analyzing droplet combustion through its high magnification images. To this end, HED addresses two key limitations of other multiscale approaches: First, it enables an all-inclusive training and prediction of images stimulated by FCNN's to generate edge maps on an image-to-image basis, and second, the pathway along which the edge is predicted is mutual to all the edge maps, thus making the successive edge maps more succinct. This section describes the formulation of the network architecture of HED and its implementation for detecting droplet edge with high fidelity.

4.1 Trimmed network architecture

In the HED method, an advanced CNN known as VGGNet[27] is used, as shown in Fig. 3.3. This is because of its broad depth, density, and multiple phases, and its exceptional performance in the ImageNet Large Scale Visual Recognition Challenge.[28] In this method, the last few layers, that is, the fifth pooling layer and the fully connected layers of this network are removed, and the side outputs in each stage are linked to the last convolutional layer, resulting in a weighted fusion output. As a result of these modifications, the trimmed network can integrate deep supervision during training, which makes the nested side outputs more natural and intuitive. This guides the model to learn features more transparently, thereby improving the accuracy.

4.2 Loss function

The optimization approach based on the stochastic gradient descent is used to train deep learning neural networks. This is accomplished by selecting a loss function[29] that calculates the error or the departure of the results obtained by the model's current state from the required output. This function is further utilized to limit the loss by updating the model weights through

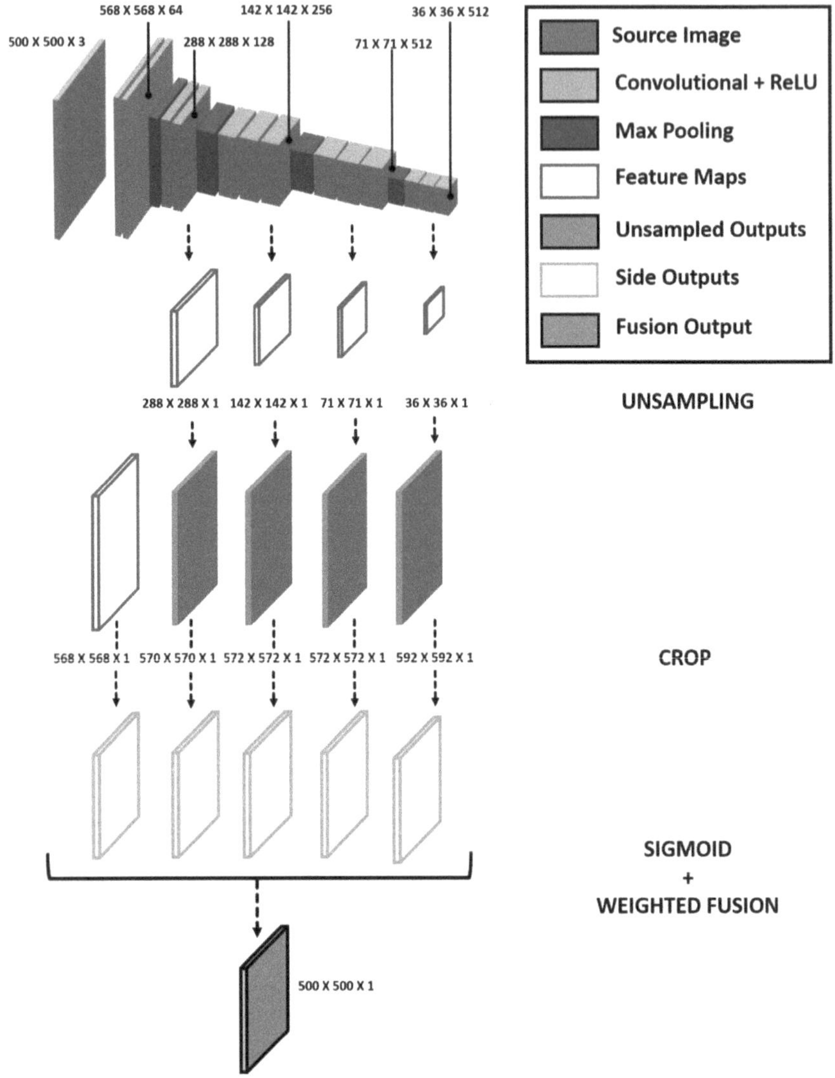

Figure 3.3 A trimmed network architecture used for edge detection. Trimmed network architecture for edge detection showing the presence of side-output layers after convolutional layers, and the application of deep-learning to each side-output layer; directing them toward the edge predictions with the desirable features that aid in detecting the droplet boundary.

computation of gradients. A weighted fusion layer joins the side-output at each level as shown in Fig. 3.3. Weighted fusion supervision is achieved by incorporating the departure of this output from the expected outcome into the loss function. Further, a weighted sum (weights are hard encoded) of the deviation from the desired end output for each side output layer is added, resulting in deep supervision while learning. In this way, the loss function integrates the deep supervision with the weight fusion supervision, thereby, causing each network layer to function as a single network, which is responsible for producing an edge map at a given scale. As a result, each side output layer contributes coherently and leads to an enhanced performance. This upgraded network was pretrained using the Berkeley Segmentation Dataset (BSDS500), which contains roughly 500 images and their manually annotated ground truth contours.

4.3 Implementation

The entire model has been implemented using the OpenCV library. OpenCV is a computer vision library that can be executed in several different programming languages, Python being the choice here. The HED model is imported via a pretrained Convolutional Architecture for Fast Feature Embedding (CAFFE) model.[30] The CAFFE is an open-source deep-learning framework that has been designed with a primary focus on speed, making it ideal for research-based implementations. This implementation using OpenCV poses some challenges as the library is not designed to run on GPU machines. Thus, to improve the speed of the model, the OpenCV library manually complied with the CUDA platform. CUDA is a parallel computing platform and programming model developed by NVIDIA for general computing on GPUs. After integrating the GPU (NVIDIA GTX 16), it takes HED 0.45 s to produce the final edge map for a single frame of dimensions 1000 pixels 800 pixels (each pixel having its own RGB value) from the droplet combustion video. This implementation is subsequently utilized to generate edge maps for each of the droplet combustion recordings (containing several frames) to obtain the droplet burn rate.

5. Results

With the large amount of data to be processed (approximately 25,000 frames per video), and the lack of ground truth for detecting droplet edge and its boundaries, a suitable option is to perform visual inspection of the results using the HED method described in Section 4. Fig. 3.4 shows the processed images resulting from the HED for the four consecutive frames results, which generates nonbinary edge maps. This means that the edges

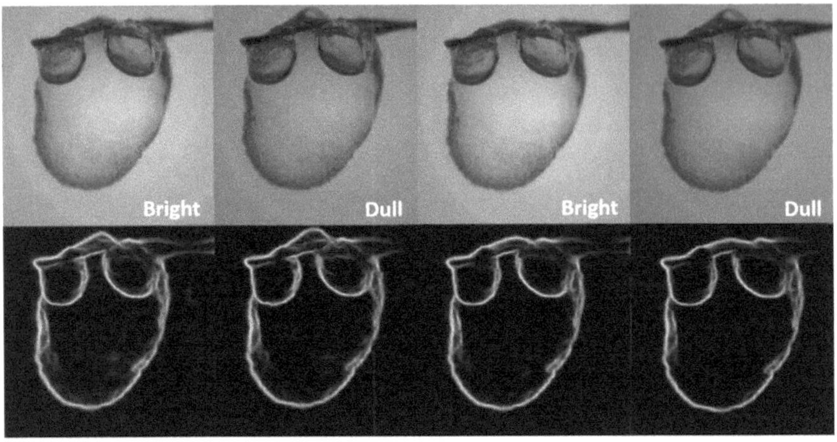

Figure 3.4 A sequence of four consecutive images obtained from the gel droplet combustion recordings. Four consecutive images obtained from the gel droplet combustion recordings: (Top row) raw images (Bottom row) corresponding processed images using the holistically nested edge detection (HED). Note the continuous droplet boundary and the absence of noise.

identified by this approach have a relative intensity, and therefore, they can be distinguished from the background noise without requiring a visual confirmation or feeding threshold value manually. Furthermore, the edges of the droplet are progressively refined and run through the entire length of the droplet boundary. The sequential process that is adapted to detect the droplet boundary and extract the geometrical parameters from the high-speed images is summarized in Fig. 3.5. The first step in this process demonstrates how the deep-supervision guided HED aids in generating multiscale dense predictions, where the noise is minimized from the image while the critical boundaries of the droplet are retained.

5.1 Color thresholding and droplet localization

The edge maps produced by HED are nonbinary and rich in resolvable turbulence. The pixels holding a lower value (less than 105) in this step are considered as noise, and therefore removed against a predetermined threshold value, which ranges between 70 and 110 for all the experimental trials. The resulting image displays a binary map with continuous and solid boundaries. The droplet is then localized using a sequence of binary bitmasks to generate a solid frame of the droplet front.

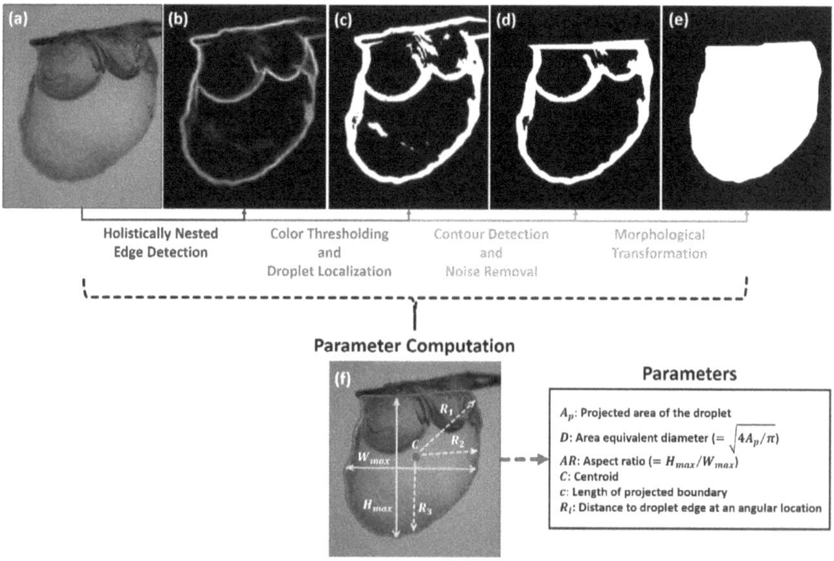

Figure 3.5 Procedure of image processing. A step-by-step procedure of image processing to extract the geometrical parameters of a burning gel fuel droplet from its raw images in the high-speed recording.

5.2 Convex Hull

During the lifetime of a burning gel droplet, an abrupt high-intensity disruption may lead to bursting of the gel droplet, as shown in Fig. 3.6. This may result in unresolved droplet boundaries due to the 3D nature of the bursting event and the variation in intensity in different sections of the image. To detect edges in such frames, the algorithm must keep a count of the number of connected components in the image and the position of their individual centroids to identify the droplet boundary. If the nonsingular components or the centroid lies outside the component area, the algorithm uses the convex hull operation to construct the droplet boundary from the given components. Convex Hull is a popular algorithm that creates a convex polygon around a given set of points. While it is a time-intensive process with a time-complexity of O (n-squared), the number of input points can be significantly reduced by choosing only the points with the minimum and the max value of the x-coordinate at each level along the y-axis. Fig. 3.6 demonstrates the resultant boundary obtained after applying convex hull operation to a bursting event with series of images, while Fig. 3.7 shows a single image to demonstrate the postprocessing operations performed to construct the droplet boundary through convex hull operation.

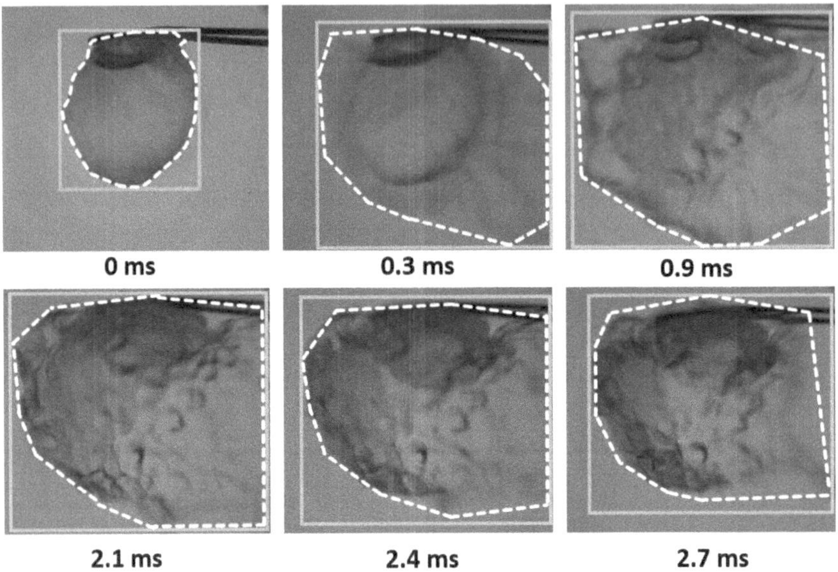

Figure 3.6 High-magnification images showing a sequential turn of events during high-intensity bursting of a gel fuel droplet. High-magnification images of a high-intensity bursting event during combustion of a gel fuel droplet. The convex polygons used to envelope the different ruptured states of the burning gel droplet are shown marked.

5.3 Morphological transformation

The droplet boundary constructed previously needs to be refined into a proper form. This is achieved by using morphological transformation, which constitutes the basic operations that are dependent on the shape of an image and are usually applied to binary images. It requires two inputs: the raw image and the structuring factor or the kernel that determines the essence of the operation. The kernel used in this study is a circular one to achieve a rounded smooth boundary. Morphological transformation is equipped with two operations, namely, opening and closing. The opening involves erosion followed by the dilation of the binary image, which is useful in removing the scattered noise from the image, while the closing is vice versa; involving dilation followed by the erosion of the binary image. It is useful in closing small holes inside the foreground objects and the small black points on the object. Fig. 3.5E depicts the uniform droplet boundary obtained by applying the morphological transformations.

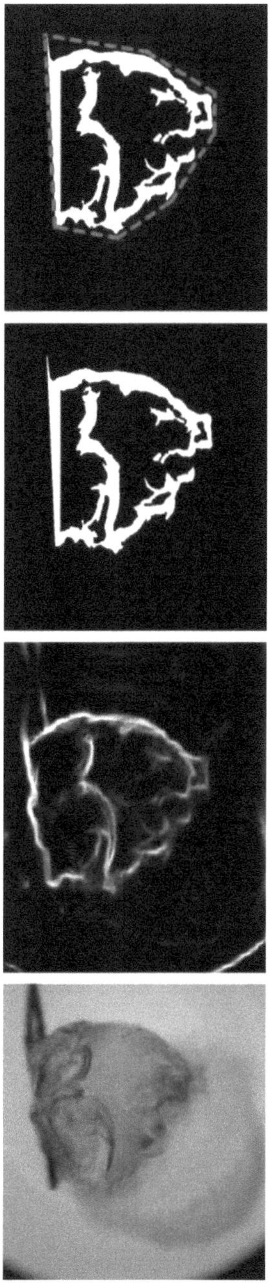

Figure 3.7 Procedure for constructing a convex polygon to detect high-intensity bursting event. A step-by-step procedure for constructing a convex polygon in the single image of a gel fuel droplet in its ruptured state.

5.4 Parameter extraction

The results as shown in Fig. 3.5F mark the exact droplet boundary, which is used to calculate the six geometrical parameters to analyze gel droplet combustion. All the parameters except the radial distance from the droplet centroid to its edge are extracted from the uniform droplet boundary (acquired from morphological transformation) using the standard Open CV library functions. A separate function is hard coded to calculate the radial distances from the centroid to the droplet edge over 1000 equal spaced angles. This function takes the location of the centroid and the droplet contour as input and runs a loop for 1000 equal spaced angles (0,/1000, 2/1000, ...) to compute the distance between the centroid and the contour coordinates along each angle using the numerical approximation. These parameters are presented in Fig. 3.5, and their significance in gel droplet combustion is detailed in Table 3.2. The time history for each of these parameters is

Table 3.2 Geometrical parameters for characterizing droplet shape variation with time.
Geometrical parameters for characterizing droplet shape variation with time and their relevance in analyzing the combustion behavior of gel fuel droplets.

Parameter	Symbol	Significance
Projected area[a]	A_p	Droplet diameter and enhancement is surface area resulting from disruption events can be determined.
Diameter[a]	D	Droplet burn rate based on the D^2−law can be determined: Calculated as the slope of $\left(D/D_o\right)^2$ versus $\left(t/D_o^2\right)$ plot, where t is the time, D is the instantaneous diameter, and D_o is the initial diameter at time $t = 0$ s.
Circumference[a]	H_{max}/W_{max}	Droplet shape deformation resulting from jetting events can be determined
Centroid	C	Centroid movement is a measure of intensity and direction of the jetting events because it determines the bulk movement of the droplet in response to vapor jetting
Radial distance to droplet edge	R_i	This length is used as an input in algorithms such as dynamic mode decomposition (DMD) to determine the key modes governing the combustion behavior of gel fuel droplets

[a]The FFT analysis of time series data of these parameters can be used to determine the dominant frequencies of droplet shape oscillations, while Continuous Wavelet Spectra (CWT) can be used to determine the time variation of dominant frequency peaks.

presented in Fig. 3.8 for a representative case of HPMC six fuel. Results from Fig. 3.8 demonstrate that the CNN-based HED can accurately capture the key effects of jetting events on the droplet geometry, which govern its gasification rate, and hence the combustion behavior.

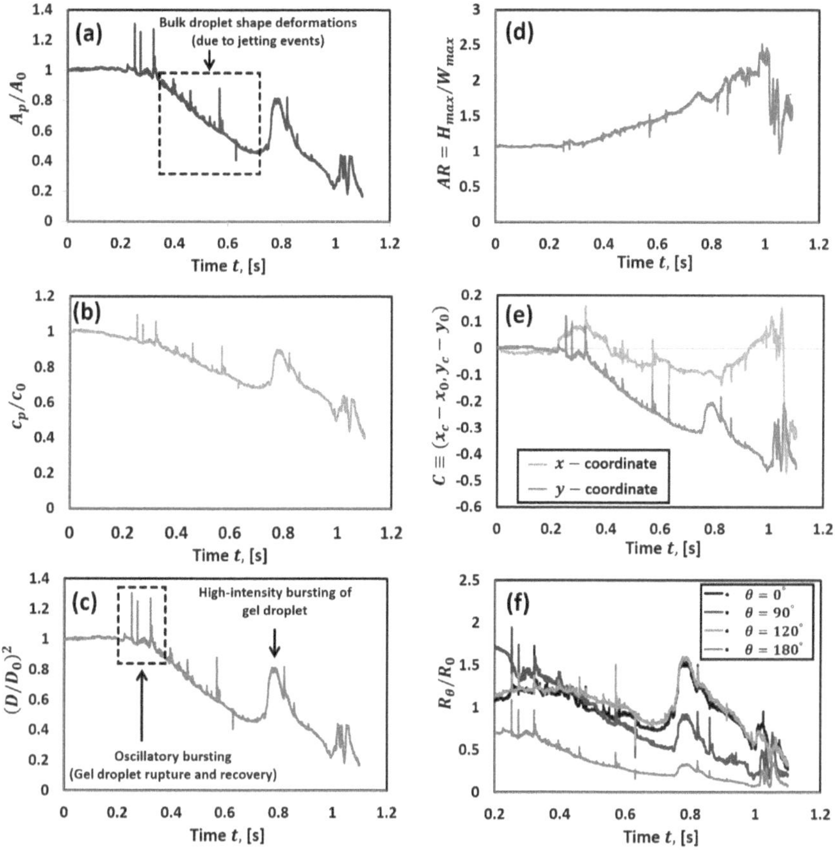

Figure 3.8 Time history of key geometrical parameters in gel droplet combustion. Temporal variation of the key geometrical parameters extracted from the high-magnification, high-speed images of a burning gel fuel droplet using the CNN-based holistically nested edge detection technique: (A) normalized projected area, (B) normalized droplet circumference, (C) squared nondimensional droplet diameter, (D) aspect ratio, (E) shift in centroid coordinates w.r.t. the initial position, and (F) normalized radial distance from instantaneous centroid to droplet edge at different angular locations. Normalization is done by the corresponding initial value i.e., at time t = 0 s.

5.5 Data verification

The extraction of geometrical parameters from the droplet boundary is accompanied by tedious mathematical computations. Due to improper logic implementation, these computations may result in wrong values for the extracted parameters. For instance, the numerical approximation method which is incorporated for radial distance might result in inaccurate results due to possible logical errors in the code. Since benchmark data are not available against which the values of parameters can be compared to determine the error, and subsequently calibrate them, it is imperative to keep a check on the correctness of the extracted data. Therefore, the data are verified by reconstructing the droplet boundary by taking a polar plot of the distances from the centroid to the droplet edge based on their angle. This is done by feeding an array of computed radial distances and their corresponding angles to the standard polar plot function from the matplotlib python library. Fig. 3.9 below demonstrates the reconstructed droplet boundary by using the extracted parameter and its close resemblance to the actual droplet.

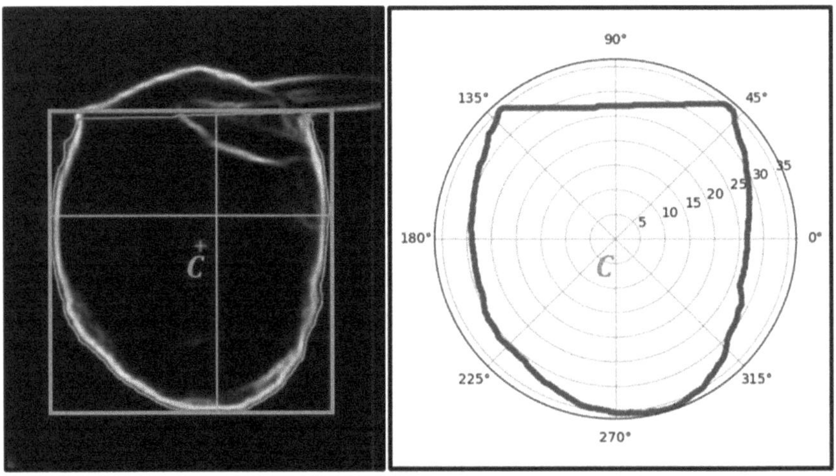

Figure 3.9 Plots showing comparison between HED obtained droplet boundary versus that obtained from polar plot function. A comparison between the droplet boundary detected using the HED (left) versus the one constructed using the polar plot function (right).

6. Concluding remarks

Currently, a significant research effort is being directed toward understanding the combustion behavior of new generation gel fuels because they are storable, nontoxic, and can offer controllability in combustors,[31–34] especially in rocket engines. Their combustion behavior is highlighted by the jetting of unreacted fuel vapors that occur throughout the droplet lifetime (depending on gellant concentration and type). The intensity and angular location of these disruptive events varies dynamically, thereby inducing a continuous change in the droplet shape as it burns. Existing methods of measuring the burn rates of such deforming droplets rely on conventional image processing methods such as structured edge, Canny Edge Detection, Otsu's thresholding,[35] to name a few, to obtain droplet diameter from a set of high-speed images. However, these methods rely on threshold input and manual feedback on an image-to-image basis, thus making the whole process time-intensive and intractable when dealing with tens of thousands of images in a single combustion recording.

This study demonstrates that deep learning-based HED technique can overcome these limitations due to automatic learning of multiscale and multilevel features in a droplet image and reconstruct the droplet boundary with high-fidelity. HED can process burning gel droplet images at a rate of 0.45 s per image, which is significantly faster compared to the CNN-based edge detection algorithms currently in use, and orders of magnitude faster compared to the conventional edge detection methods. This can help reduce the computational time significantly when dealing with ultrahigh-speed videos, as encountered during experimentation with burning gel fuel droplets. The nonbinary edge maps produced by the HED are mostly discontinuous, and rich in turbulence, and therefore, require restructuring. Therefore, a multistaged postprocessing algorithm is developed which removes the noise and pieces together the entire droplet to produce one continuous boundary. To this end, the HED technique combined with a postprocessing algorithm developed in the study is computationally efficient, and extremely effective in detecting the droplet edges reliably (and hence the droplet shape to be reconstructed). This allows the droplet burn rates to be determined with extremely high accuracy even under disruptive combustion conditions, which are often encountered in practical combustion devices.

References

1. Maurice LQ, Lander H, Edwards T, Harrison WE. Advanced aviation fuels: a look ahead via a historical perspective. *Fuel.* 2001;80(5):747–756. https://doi.org/10.1016/S0016-2361(00)00142-3.
2. Miglani A, Nandagopalan P, John J, Baek SW. Oscillatory bursting of gel fuel droplets in a reacting environment. *Sci Rep.* 2017;7(1). https://doi.org/10.1038/s41598-017-03221-x, 3088-3088.
3. Natan B, Rahimi S. The status OF gel propellants IN year 2000. *Int J Energ Mater Chem Propuls.* 2002;5(1–6):172–194. https://doi.org/10.1615/intjenergeticmaterialschemprop.v5.i1-6.200.
4. Allan BD. *Gelled Monomethylhydrazine Thixotropic Fuel - Patent US-4039360*; September 1977. https://patents.google.com/patent/US4039360.
5. John J, Nandagopalan P, Baek SW, Miglani A. Rheology of solid-like ethanol fuel for hybrid rockets: effect of type and concentration of gellants. *Fuel.* 2017;209:96–108. https://doi.org/10.1016/J.FUEL.2017.06.124.
6. Nandagopalan P, John J, Baek SW, Miglani A, Ardhianto K. Shear-flow rheology and viscoelastic instabilities of ethanol gel fuels. *Exp Therm Fluid Sci.* 2018;99:181–189. https://doi.org/10.1016/J.EXPTHERMFLUSCI.2018.07.024.
7. Tarpley Jr WB. *Thixotropic Liquid Propellant Compositions With Solid Storage Characteristics-Patent US-3470040*; September 1969. https://patents.google.com/patent/US3470040.
8. Miglani A, Nandagopalan P, John J, Baek S. *Disruptive Combustion Behavior of Gelled Ethanol Fuel Droplets.* 2016.
9. Arnold R, Anderson W. Droplet burning of JP-8/silica gels. In: *48th AIAA Aerospace Sciences Meeting Including the New Horizons Forum and Aerospace Exposition.* American Institute of Aeronautics and Astronautics; 2010. https://doi.org/10.2514/6.2010-421.
10. Miglani A, Basu S, Kumar R. Insight into instabilities in burning droplets. *Phys Fluids.* 2014;26(3). https://doi.org/10.1063/1.4866866, 32101-32101.
11. Solomon Y, DeFini SJ, Pourpoint TL, Anderson WE. Gelled monomethyl hydrazine hypergolic droplet investigation. *J Propul Power.* 2012;29(1):79–86. https://doi.org/10.2514/1.B34634.
12. Miglani A, Basu S. Effect of particle concentration on shape deformation and secondary atomization characteristics of a burning nanotitania dispersion droplet. *J Heat Tran.* 2015;137(10). https://doi.org/10.1115/1.4030394.
13. Miglani A, Basu S, Kumar R. Suppression of instabilities in burning droplets using preferential acoustic perturbations. *Combust Flame.* 2014;161(12):3181–3190. https://doi.org/10.1016/J.COMBUSTFLAME.2014.06.010.
14. Dollár P, Zitnick CL. Fast edge detection using structured forests. *IEEE Trans Pattern Anal Mach Intell.* 2015;37(8):1558–1570. https://doi.org/10.1109/TPAMI.2014.2377715.
15. Canny JA. Computational approach to edge detection. *IEEE Trans Pattern Anal Mach Intell.* 1986;6:679–698. https://doi.org/10.1109/TPAMI.1986.4767851. PAMI-8.
16. Xie S, Tu Z. Holistically-nested edge detection. In: *2015 IEEE International Conference on Computer Vision (ICCV).* 2015:1395–1403. https://doi.org/10.1109/ICCV.2015.164.
17. Ye QZ. A fast algorithm for convex hull extraction in 2D images. *Pattern Recogn Lett.* 1995;16(5):531–537. https://doi.org/10.1016/0167-8655(95)00122-W.
18. Basu S, Miglani A. Combustion and heat transfer characteristics of nanofluid fuel droplets: a short review. *Int J Heat Mass Tran.* 2016;96:482–503. https://doi.org/10.1016/J.IJHEATMASSTRANSFER.2016.01.053.

19. Miglani A, Basu S. Coupled mechanisms of precipitation and atomization in burning nanofluid fuel droplets. *Sci Rep*. 2015;5(1). https://doi.org/10.1038/srep15008, 15008-15008.
20. Kittler J. On the accuracy of the Sobel edge detector. *Image Vis Comput*. 1983;1(1): 37–42. https://doi.org/10.1016/0262-8856(83)90006-9.
21. Bertasius G, Shi J, Torresani L. DeepEdge: a multi-scale bifurcated deep network for top-down contour detection. In: *2015 IEEE Conference on Computer Vision and Pattern Recognition (CVPR)*. 2015:4380–4389. https://doi.org/10.1109/CVPR.2015.7299067.
22. Ganin Y, Lempitsky V. *$$N4$$-Fields: Neural Network Nearest Neighbor Fields for Image Transforms BT - Computer Vision – ACCV 2014*. 2015:536–551.
23. Hwang J-J, Liu T-L. *Pixel-wise Deep Learning for Contour Detection*. April 2015.
24. Shen W, Wang X, Wang Y, Bai X, Zhang Z. DeepContour: a deep convolutional feature learned by positive-sharing loss for contour detection. In: *2015 IEEE Conference on Computer Vision and Pattern Recognition (CVPR)*. 2015:3982–3991. https://doi.org/10.1109/CVPR.2015.7299024.
25. Sokolova M, Japkowicz N, Szpakowicz S. Beyond accuracy, F-score and ROC: a family of discriminant measures for performance evaluation BT - AI 2006. *Adv Artif Intell*. 2006:1015–1021.
26. Martin DR, Fowlkes CC, Malik J. Learning to detect natural image boundaries using local brightness, color, and texture cues. *IEEE Trans Pattern Anal Mach Intell*. 2004; 26(5):530–549. https://doi.org/10.1109/TPAMI.2004.1273918.
27. Simonyan K, Zisserman A. *Very Deep Convolutional Networks for Large-Scale Image Recognition*. arXiv 14091556; September 2014.
28. Russakovsky O, Deng J, Su H, et al. ImageNet large scale visual recognition challenge. *Int J Comput Vis*. 2015;115(3):211–252. https://doi.org/10.1007/s11263-015-0816-y.
29. Zung J, Tartavull I, Seung H. *An Error Detection and Correction Framework for Connectomics*. August 2017.
30. Long J, Shelhamer E, Darrell T. Fully convolutional networks for semantic segmentation. In: *2015 IEEE Conference on Computer Vision and Pattern Recognition (CVPR)*. 2015: 3431–3440. https://doi.org/10.1109/CVPR.2015.7298965.
31. Idahosa U, Basu S, Miglani A. System level analysis of acoustically forced nonpremixed swirling flames. *J Therm Sci Eng Appl*. 2014;6(3). https://doi.org/10.1115/1.4027297.
32. Idahosa U, Santhosh R, Miglani A, Basu S. Response dynamics of recirculation structures in coaxial nonpremixed swirl-stabilized flames subjected to acoustic forcing. *J Therm Sci Eng Appl*. 2015;8(1). https://doi.org/10.1115/1.4030728.
33. Santhosh R, Miglani A, Basu S. Transition and acoustic response of recirculation structures in an unconfined co-axial isothermal swirling flow. *Phys Fluids*. 2013;25(8). https://doi.org/10.1063/1.4817665, 83603-83603.
34. Santhosh R, Miglani A, Basu S. Transition in vortex breakdown modes in a coaxial isothermal unconfined swirling jet. *Phys Fluids*. 2014;26(4). https://doi.org/10.1063/1.4870016, 43601-43601.
35. Goh TY, Basah SN, Yazid H, Aziz Safar MJ, Ahmad Saad FS. Performance analysis of image thresholding: Otsu technique. *Measurement*. 2018;114:298–307. https://doi.org/10.1016/J.MEASUREMENT.2017.09.052.

CHAPTER FOUR

Deep learning-based methods for detecting surface defects in steel plates

Sakshee Patil[2], Ankur Miglani[2], Pavan Kumar Kankar[1], Debanik Roy[3]

[1]System Dynamics Lab, Department of Mechanical Engineering, Indian Institute of Technology, Indore, Madhya Pradesh, India
[2]Microfluidics and Droplet Dynamics Lab, Department of Mechanical Engineering, Indian Institute of Technology, Indore, Madhya Pradesh, India
[3]Bhabha Atomic Research Centre & Homi Bhabha National Institute, Department of Atomic Energy, Mumbai, Maharashtra, India

Contents

Smart Electrical and Mechanical Systems
ISBN: 978-0-323-90789-7
https://doi.org/10.1016/B978-0-323-90789-7.00001-4

1. Introduction

Steel is primarily an alloy of iron and carbon whose mechanical properties are primarily influenced by the percentage of carbon that it contains. Steels are highly elastic, ductile, and malleable materials that can be welded and forged, thereby making them an ideal material for construction. However, several defects may develop in the material during the process of steel manufacturing, fabrication, storage, or during operation in practice. The commonly seen bulk defects in steel are porosity, gas holes, shrinkage, shell, slivers, nonmetallic inclusions, centerline segregation cracking. The key surface defects include the presence of scratches, notches, and corrosion-induced pitting marks, which may act as stress concentration points and lead to the initiation of cracks in an otherwise healthy material. The presence of such defects may reduce the material strength significantly, thereby affecting the predictability and reliability of steel-based products. This apparent issue makes the problem of steel defect detection important. The idea of automating the defect detection process has attracted significant attention of the steel production industry in the past decade.[1,2] The first step in the automated steel surface inspection system is to prepare a robust database of high-resolution images that are representative of different types of surface defects and their severity. The next step involves the extraction of features from these images that helps in categorizing the defects based on their individual characteristics. Subsequently, the labels are assigned to the different classes of characteristics that are retrieved. Once this dataset is optimized, accurate, and varied to the extent possible, the defect classification and segmentation are explored; wherein a deep learning model is trained over the dataset for defect detection and classification. These steps form the foundation of any automated steel surface inspection system and have been explored over the past decade in varied depth.[3–5] Previously, semiautomation has been explored, wherein traditional machine vision[6,7] was used to aid manual visual inspection.[8,9] However, low confidence despite the involvement of large manpower made the manual part of the process highly time-consuming and subjective. More recently, several methods have been proposed which rely on first extracting the features manually, and then passing them through the machine learning classifiers to categorize defects.[10] A key limitation of these methods is their dependence on the quality and quantity of the training dataset, which is generally difficult to construct; because of the time, cost, and human resources required to collect data and construct a robust database. Consequently, the limited training data ends up being generalized for multiple types of testing scenes, which leads to low classification accuracies. Low classification accuracy is a major constraint and makes these methods ineffective for the steel industry, which must meet the large production demand that is defect-free. To enhance the detection and classification accuracy of surface defects, previous studies have proposed convolutional neural networks (CNNs)-based machine learning approaches.[11,12] In these approaches, the two deep learning algorithms are commonly utilized, namely, image classification and object detection. The first method[13–15] identifies complete images without specifying the defect location in the images. The second method is further divided into object detection and segmentation. Object detection[16] refers to drawing a bounding box around each detected defect. Even though the bounding boxes provide a rough estimate about the location and tentative size of the defect, it is prone to detecting the extra defect-free region surrounding the defect. In this regard, this localization technique is not too effective for detecting surface defects since it provides a vague estimate of the defect at best. Semantic segmentation of the object[17,18] can be used to address the challenge of demarcating defects from the defect-free region (as encountered by bounding boxes) accurately by producing pixel-perfect masks. These pixel-perfect masks are obtained by hot encoding the images into masks corresponding to each distinct class of the defect.

Deep learning methods rely on extensive and robust training data to enable them to generalize well. Since each new image brings a level of newness in terms of the features, the model's performance improves as the dataset of the manually annotated image examples grows. Constructing such a dataset is a time-intensive operation best left to the trained staff. Therefore, the dataset used in this is the open source Severstal Steel Defect Dataset.[19] This dataset consists of 6666 labeled images spread across four types of surface defects in steel plates. The ground truth segmentation masks are also provided in the form of run-length encoded pixels. Literature search reveals that one of the most widely used image segmentation models, namely, the original U-Net architecture, produces a dice coefficient less than 0.5 for this dataset, which is insufficient to justify the use of automation over manual inspection for defect detection. To overcome this discrepancy of needing automation but having poor automation results, the twofold objective of this study is to first maximize the dice score while localizing the defect present in the image, and second, to maximize the multilabel probability estimates as summarized in Fig. 4.1.

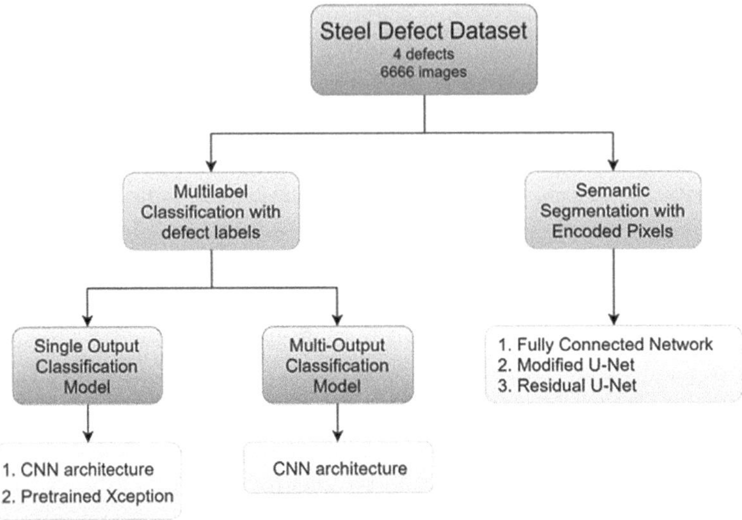

Figure 4.1 Flowchart showing the methodology adapted for detecting surface defects in steel plates using deep learning-based multilabel image segmentation and classification techniques.

In this chapter, multilabel classification and semantic segmentation are explored separately for defect detection in steel plates (see Fig. 4.1). Semantic segmentation can generate pixel-wise masks of defects while classifying them under a class label. Therefore, a baseline fully connected network is trained as a baseline and the pertinent modifications are made to the well-known U-Net architecture to optimize the dice scores further. To validate the classification results of the segmentation models, both single and multi-output CNN-based architectures are explored. Even though the classification does not indicate the exact location of the defect, it can aid in manual localization because of its high accuracy.

2. Dataset

The Severstal Steel Defect dataset is used in this study, which consists of high-resolution images of steel sheets with four fault classes and areas. The dataset has a total of 6666 grayscale images, each with a resolution of 256 × 1600 pixels. Further, each image has a corresponding defect label and a run-length encoded ground truth segmentation map.

Defects are categorized into four classes: type 1, type 2, type 3, and type 4. There are 897 photos with type 1 defects, 247 images with type 2 defects,

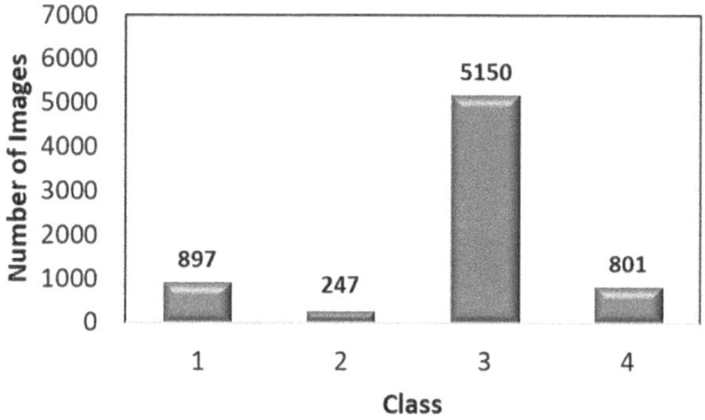

Figure 4.2 Bar graph showing imbalance between the number of images in each defect class.

5150 images with type 3 defects, and 801 images with type 4 defects. The imbalance in the dataset is apparent from Fig. 4.2, which shows the class distribution. Out of 6666 images, 6239 images contain only one defect class, 425 images contain two defect classes, and only two images have three defect classes. Two samples for each defect type from the dataset are shown in Fig. 4.3. It is observed that many tiny size regions are present in type 1 defect type 1 photos, whereas multiple medium-size regions are present in type 4 defects. Multiple medium-sized regions are also visible type 3 defects. While the images of defect type 2 and type 3 share some regional characteristics, they are not identical.

3. Metrics

3.1 Dice coefficient

This metric is used to determine the similarity index of two samples. It can be used to compare the pixel-wise overlap between a predicted mask and its target. The dice coefficient is calculated as 2 times the area of overlap divided by the total number of pixels in both images. This equation is represented visually in Fig. 4.4.

The overlap region indicates genuine positives and defines how the two elements are matched with each other (TP). The difference between the ground truth and TP causes false-negatives. The rest of the union is made up of false-positives. True-negatives are another element of the image that are not predicted by segmentation.

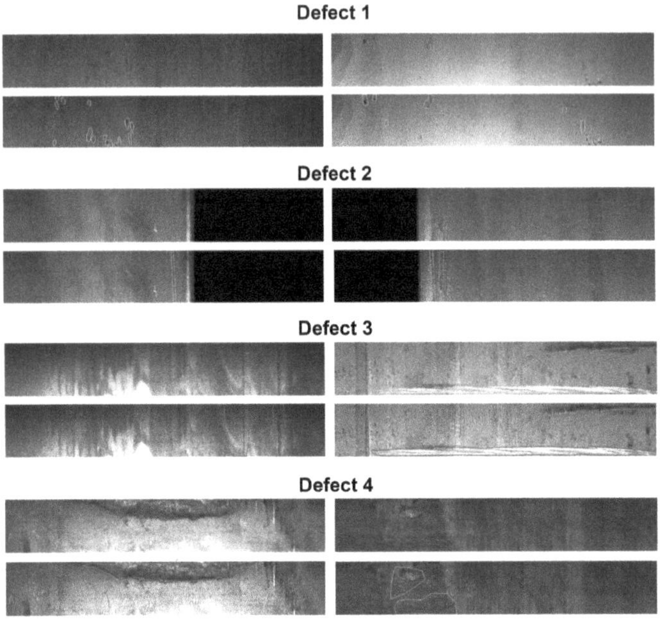

Figure 4.3 Representative images of surface defects in steel plates provided in the Severstal Steel Defect Detection Dataset.[19] Top row in each defect shows raw images. The defect detected after processing are shown marked in bottom images. *From Severstal: Steel Defect Detection; n.d. Retrieved July 2, 2021, from https://www.kaggle.com/c/severstal-steel-defect-detection/data.*

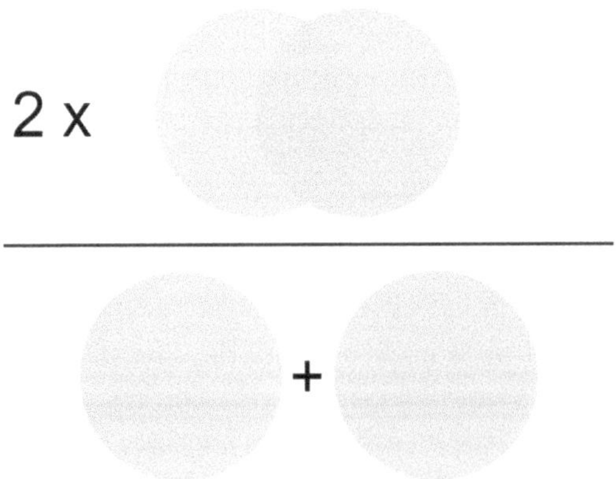

Figure 4.4 Visualization of the equation of the dice coefficient.

3.2 Precision

Precision is a measure of quality and defines how many true positives are correct. An algorithm with a higher precision score will result in more relevant results than irrelevant ones.

3.3 Recall

The recall is a measure of the quantity and defines the proportion of actual positives that are identified correctly. An algorithm with a high recall value will result in the most relevant results whether the irrelevant ones are returned or not.

3.4 Mean IoU

The Jaccard index, which is commonly known as the Intersection over Union (IoU), is a metric that quantifies the percentage overlap between the ground truth and the predicted mask and is like the dice coefficient The mean IoU value is obtained by dividing the total number of pixels present across both the masks by the number of pixels shared between the target and prediction masks.

4. Methodology

4.1 Multilabel classification

Both single and multioutput models are trained to check if the overall accuracy of the single output model is swayed due to the high probability estimate of any one class type; because of the class imbalance. Fig. 4.5 shows the CNN design used for training the single and multioutput classifiers from the multilabel classification.

4.2 Single output models

A single output deep learning model is like the normal regression model wherein a single output is predicted for each sample. In this study, two separate single output classifiers, namely, selfdesigned CNN, and the pretrained Xception are trained for the purpose of image classification.

4.2.1 CNN model

The Keras framework is utilized to design a multilabel CNN Classifier with ReLU activations, necessary dropout, batch normalization, and Maxpool layers. Dropouts help to "switch off" the neurons and their connections

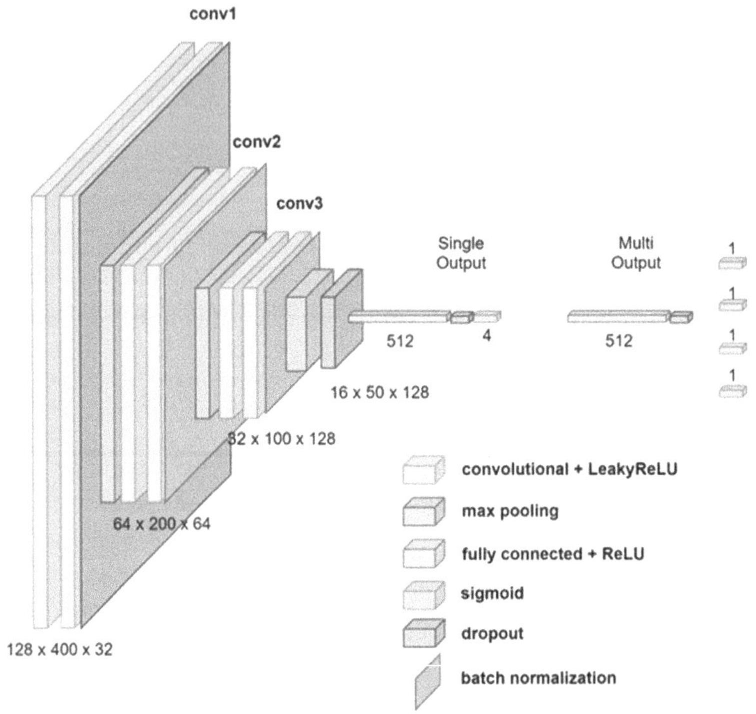

Figure 4.5 CNN classifier with both single and multioutput model visualization for surface defect detection in steel plates.

during a pass of the neural network and help in enhancing the model learning by avoiding overfitting. Batch normalization aids in normalizing or scaling the input of the previous layers. It is another regularization technique that helps in avoiding overfitting. Maxpool layers are added to reduce the dimension of the input which in turn reduces the number of learnable parameters. It also helps in creating a robust model that is immune to the variations in the position of the features (present in the input image). Besides the network structure, the Keras Image Generator is used to augment images on the fly with random horizontal and vertical flips, rotation, and brightness adjustment to tackle with the data imbalance as well as to enhance generalization. Augmentations such as cropping and zooming are not used because in many images the small defects are either at one side or only at the edges of the image. Zooming and cropping results in "no defect" images being marked with a defect label, thereby compromising the integrity of the data. The flow_from_dataframe method is used to input the batches of data during training. The images are resized from the initial 256× 1600 pixels to 128 × 400 pixels to reduce the computation time and allow a batch size of 32 to be passed for training during a single processing. A 85-15 split was made to create the validation dataset. To handle data imbalance, an adaptive learning rate method, namely, the RMSProp Optimizer is used over the learning rate schedulers like the stochastic gradient descent . Learning rate schedules aim to adjust the learning rate during training time and the adjustment depends on a predefined schedule, which is heavily influenced by the type of model and the problem. This presents an unnecessary challenge without significant enhancement of results. Among the available adaptive optimizers, Adagrad and Adadelta are preferred for only large-scale networks, and therefore, not suitable for this study. Instead, RMSProp is chosen because it shows better results compared to the Adam optimizer. Finally, the data are flattened before feeding it to the final output dense layer. This layer of the architecture is assigned a value of four neurons corresponding to each defect class with a sigmoid activation.

4.2.2 Xception model

The pretrained Xception Model with imagenet weights is used by removing the last fully connected layer of the network. For model compilation, the Adam optimizer is used for its adaptive learning rate involving a binary cross-entropy loss and accuracy metrics. Since the problem deals with multi-label classification, the CNN output vector should not be affected by other component values, that is, the probability outputs should not result in all classes for each image. To avoid such an occurrence, the binary cross-entropy loss is used instead of SoftMax loss. The final dense layer has four neurons for each defect activated by a sigmoid function.

4.3 Multioutput CNN model

The feature learning network used in the multioutput CNN model is the same as that of the single output CNN classifier discussed earlier. The only difference between their architectures is how the output layer(s) are defined. Instead of one dense layer in the single output model, four different single neuron dense output layers are defined in the multioutput model corresponding to each defect type.

4.4 Image segmentation

Three image segmentation architectures are explored in this study:

1. Fully connected network (FCN)

2. Hyper-tuned modified U-Net
3. Residual U-Net

It is demonstrated that the highest accuracy is achieved with the hyper tuned modified U-Net architecture with a dice coefficient of 0.688. In each case, the input is a steel sheet image (256, 1600, 1), and the output is a segmented image of the same size with the area constituting the defect marked therein. Fig. 4.6 shows the general training pattern followed by each of the segmentation models.

4.5 Fully connected network

The Fully Connected Network (FCN) is the most prominent CNN-based semantic segmentation algorithm used to accomplish pixel-wise prediction in an end-to-end fashion, wherein the fully connected layers are converted to 1×1 convolutional layers. A fully convolutional net is defined as a baseline for segmentation that combines feature hierarchy layers and improves output spatial precision. First, the conv layers are defined, wherein each layer of data is a 3D array of size $h \times w \times l$, with h and w representing the height and width of the image and l representing the grayscale dimension. Deconvolutional layers are then applied to up sample the image back to its original dimensions. Skips are included to combine the lower layers with the final output layers. This is done to take advantage of the finer strides of the initial layers because they observe fewer pixels. This helps the model to take into consideration the overall structure by fusing the coarser features of the final layers with the finer outputs of the initial layers to make local predictions. The idea behind convolution, deconvolution, and skip connections is visually represented in Fig. 4.7.

4.6 Hyper-tuned modified U-Net

The proposed model is inspired by the original U-Net network model. Modifications such as the LeakyReLU activation, pertinent batch normalizations, up sampling, and 20% dropout have been incorporated to hypertune this architecture to fulfill the requirements of the present study. The architecture is completely convolutional and corresponds to an autoencoder with concatenation skip connections (not residual) between blocks of the same spatial size as shown in Fig. 4.8.

The architecture can be divided into an encoder-decoder path. The encoder is made up of four 3×3 convolutions that are applied repeatedly. A batch normalization and LeakyReLU activation follow each conv. The

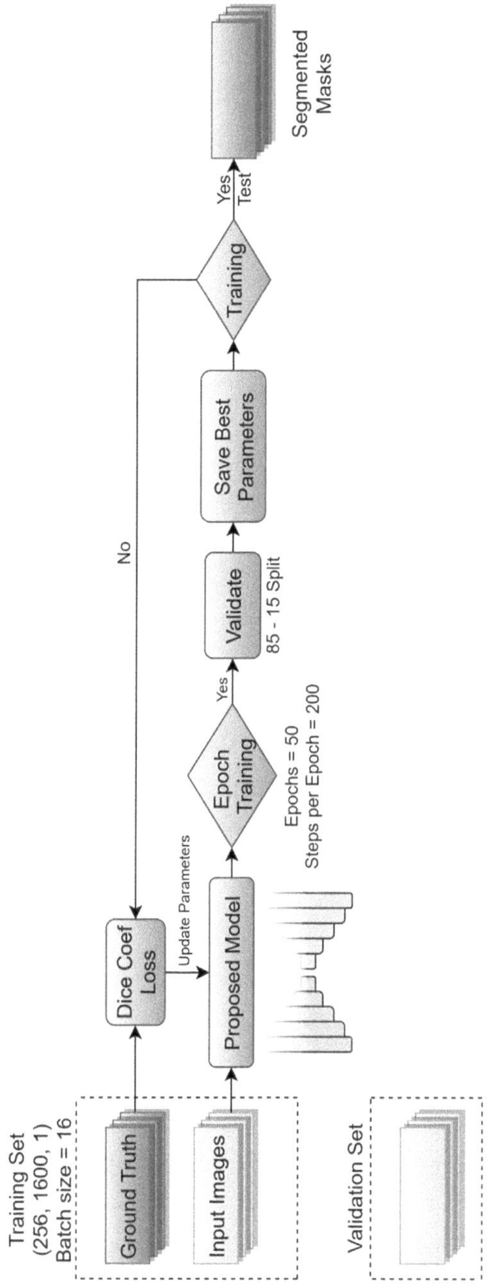

Figure 4.6 Training flowchart for the image segmentation.

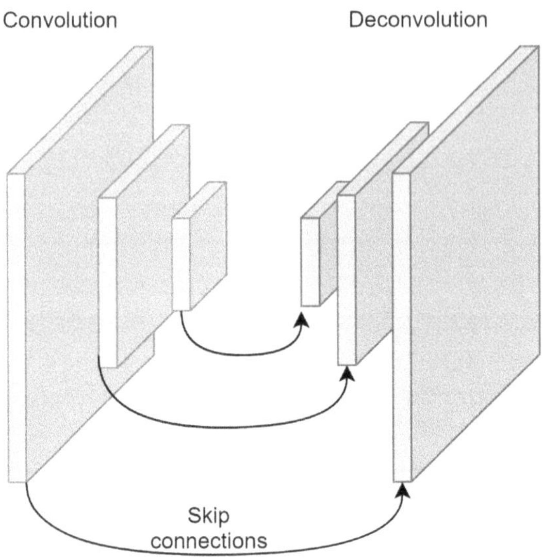

Figure 4.7 Encoder-Decoder model with skip connections.

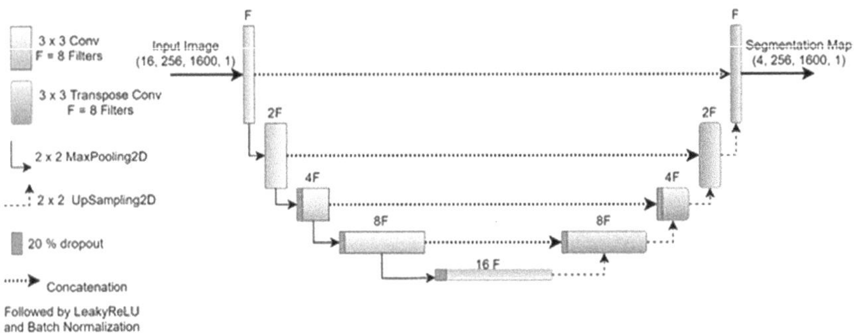

Figure 4.8 The hyper-tuned modified U-Net architecture.

spatial dimensions are then reduced using a 2 × 2 max pooling operation. The number of feature channels is doubled while halving the spatial dimensions at each of the down sampling steps. The decoder consists of a feature map up sampling followed by a 2 × 2 transpose convolution, which reduces the number of feature channels in half. Concatenation with the encoder's matching feature map and a 3 × 3 convolutional map is developed, both of which are followed by a LeakyReLU activation. A 1 × 1 convolution is utilized to map the channels to the desired four classes of defects in the final layer. Batch normalization and dropout layers are added for regularization and to take care of overfitting since the trained network is quite deep.

The original U-Net design uses ReLU activations for convolutional layers. Since activation functions introduce nonlinearity into the network, they are critical for artificial neural networks to uncover and make sense of the complicated and nonlinear correlations between the input and the output. Their goal is to turn a neural network's input signal layer into an output signal. In deep neural networks, ReLU is the most popular activation function. However, one of its shortcomings of the ReLU rectifier is that several gradients can become unstable during training and "die," thus causing the weight to update in such a way that it never activates on another data point. As a result, the ReLU can result in "dead neurons."

The gradient will be zero for activations in the ReLU zone ($x < 0$), and the weights 0 will not be fine-tuned during descent. As a result, the neurons in that state will stop responding to the changes in the input. Consequently, nothing will change simply because the gradient is zero; this is known as the "dying ReLU" issue. To overcome the downsides of ReLU nonlinearity, experimentation with other activation functions such as LeakyReLU, PReLU, Swish, and Mish is performed. The LeakyReLU activation with an alpha value of 0.1 proved to be the most effective. The difference between ReLU and LeakuReLU can be best understood from Fig. 4.9.

4.7 Residual U-Net architecture

To further optimize the results, deeper architectures like the Residual U-Net are investigated. However, increasing the depth may cause the problem of vanishing gradient in which the weight changes become insignificant, and the training process is damped. Stacking the convolutional layers hinders the generalization ability of the model.

To deal with this issue, the concept of residual blocks is introduced in the network. As shown in Fig. 4.10, a traditional neural net block learns to map the identity function $x \rightarrow f(x)$. However, a residual neural net block learns to map $x \rightarrow R(x)$, where $R(x)$ is the residual of $f(x)$, that is, $f(x) - x$. Another advantage of these blocks is that the inputs can propagate faster through the residual shortcut connections.

This residual U-Net architecture is inspired by deep residual learning and the U-Net. Due to the limited GPU memory, the depth and the hyperparameters of the architecture are slightly modified from the original.

The three components of Residual U-Net are as follows:

1. Encoder
2. Bridge
3. Decoder

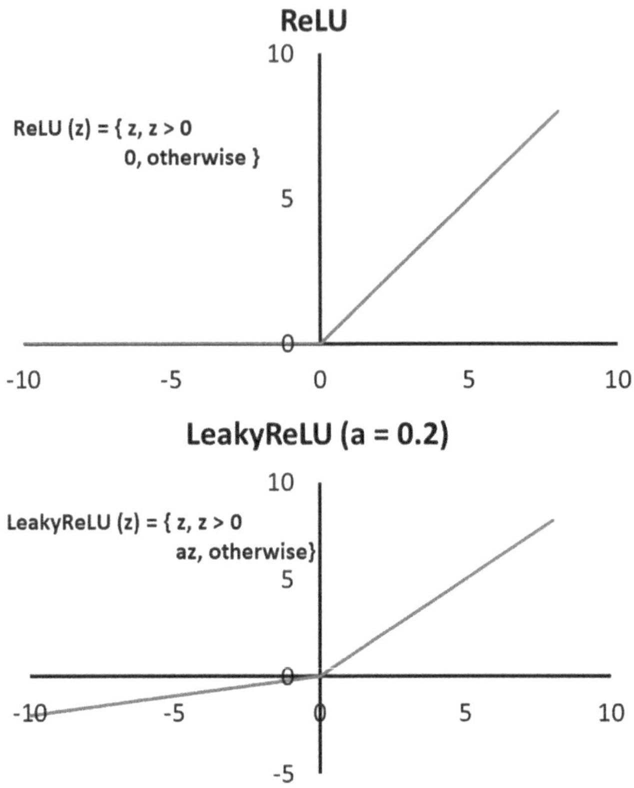

Figure 4.9 Difference between the ReLU (top), and the Leaky-ReLU (bottom) activation functions.

In the encoder units, a stride of two is applied to the first convolutional block to reduce the feature map to half its size, and the stride functionality is used instead of a pooling operation. Before each decoding unit, an up sampling and concatenation layer is added. These layers acquire the feature maps from the lower level from the appropriate encoding units. At last, a 1×1 convolution is applied with a sigmoid activation to obtain the desired four channel segmentation map. The complete Residual U-Net architecture is shown in Fig. 4.11. Note that the number of individual encoder-decoder blocks is reduced in the figure for compact visualization.

5. Results and inference

The accuracy and loss curves are plotted across epochs for the trained classification models, which allow the training progress of the networks to be

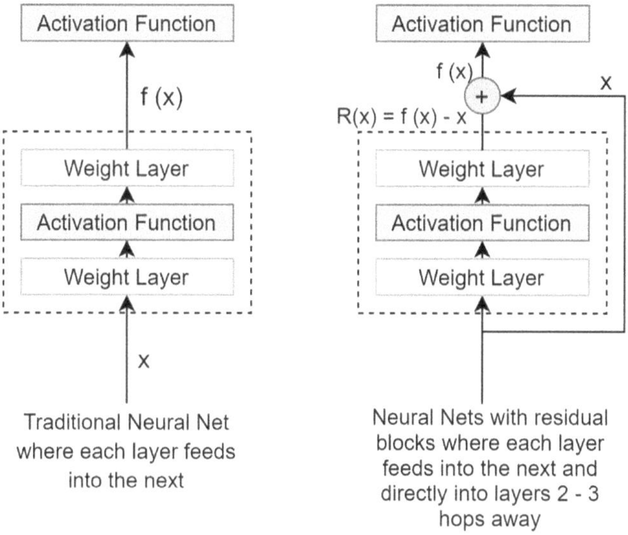

Figure 4.10 Schematic representation of the residual block.

observed. These plots also help in understanding how well a given model generalizes. If the gap between the validation line and the training line increases with time, with the validation scores lagging, it can be inferred that the model is overfitting the training data. This means that the model is getting good at learning the intricacies of the training data but at the cost of losing its generalizability over the new data. Similarly, if the training scores start lagging, the model is said to be underfitted. From Fig. 4.12, it is apparent that all the plots exhibit a good generalization potential with a steady increase in the prediction accuracy.

5.1 Multilabel classification

As discussed previously, two single output classifiers are trained for 50 epochs with an early stopping callback. Their validation accuracy is monitored with a patience of 10 for the callback. Fig. 4.12 shows the binary cross-entropy loss and the accuracy plotted as a function of epochs of the trained models.

5.2 Single output classifiers

It can be observed from Fig. 4.12 that the CNN classifier is trained for 36 epochs and indicates that a maximum accuracy of 92.045% was achieved at the 26th epoch. The batch size used for training is 32. The gradually decreasing trend of the loss curve and corresponding increasing trend of the accuracy curve implies that the model is learning over each epoch.

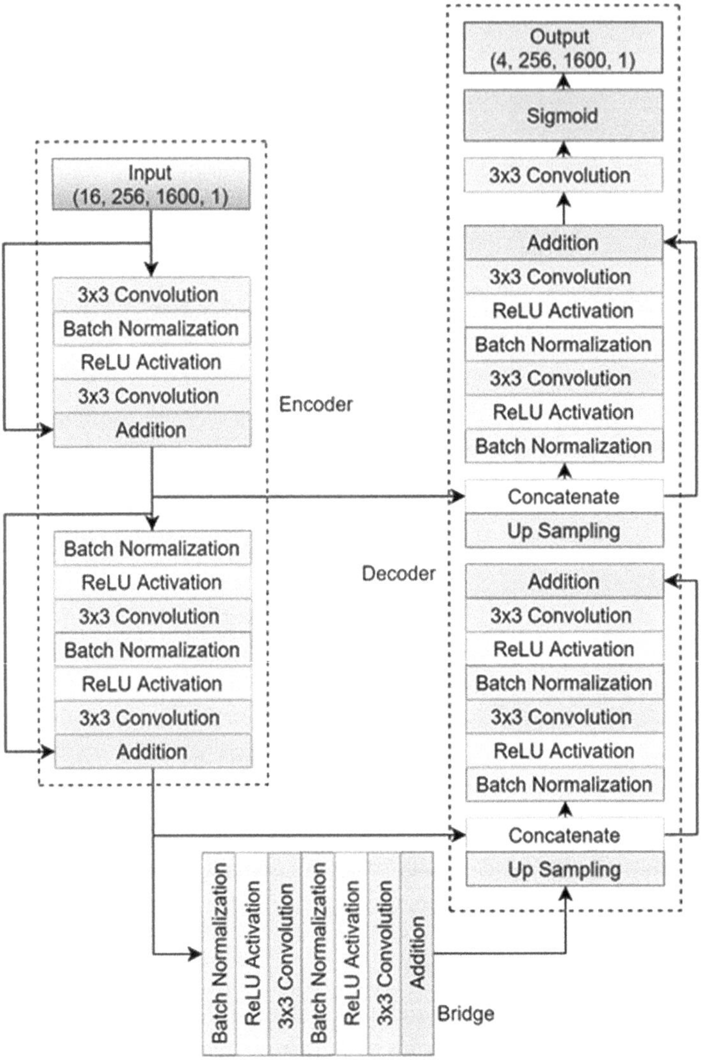

Figure 4.11 Schematic representation of the Residual U-Net architecture used for training.

Fig. 4.13 provides an estimate of the training progress of the Xception classifier. It is evident that the Xception classifier is trained for 39 epochs and that the maximum accuracy of 96.513% is achieved at the 29th epoch. The corresponding loss value is 0.0672. Note that the Xception classifier surpasses the accuracy obtained using the CNN classifier at the third epoch itself, thus showcasing the superiority of the transfer learning approach. Furthermore, the validation curves show fewer fluctuations, which implies a more stable learning curve compared to the CNN classifier.

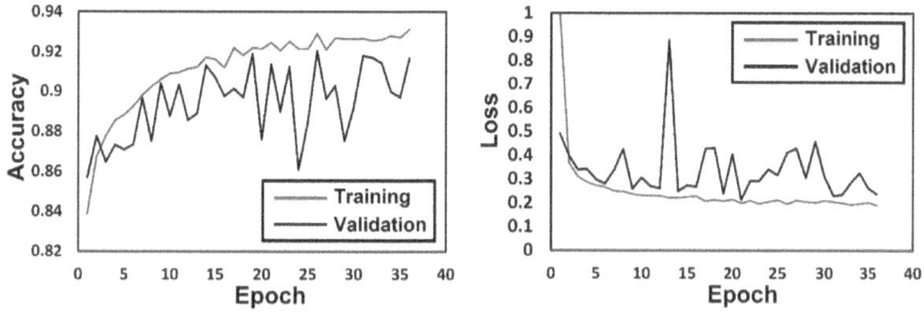

Figure 4.12 The accuracy (left) and loss (right) curves for a single output CNN classifier.

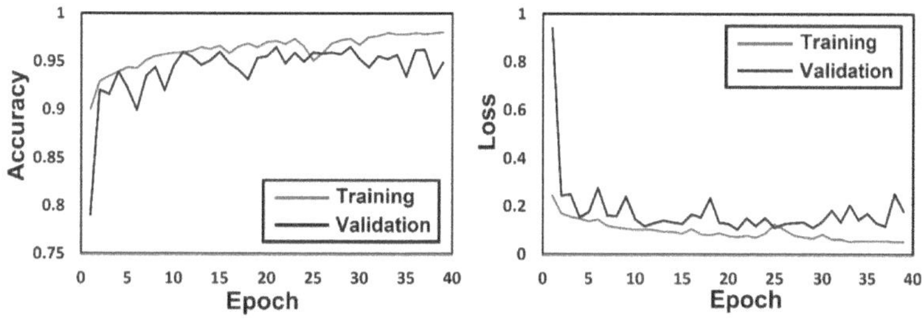

Figure 4.13 Accuracy (left) and loss (right) curves of pretrained Xception classifier.

5.3 Multi-output CNN classifier

Table 4.1 summarizes the evaluation metric of the multi-output CNN classifier. It is observed that in the overall accuracy of 92.045% for the single output classifier, the classification of defect two contributes the most. Despite the lowest number of images in defect two and the highest number of images in defect 3, their accuracies are the highest and lowest, respectively. The way this observation can be understood is to use the reweighting method from the scikit-learn library to estimate the individual class weights for the unbalanced dataset. In short, the lower the data in a certain class, the higher weight it gets. This influences the loss function by assigning relatively higher costs to the examples from the minority classes. However, this method did not show any noticeable increase in the scores. Therefore, the high validation accuracy of the minority class can be attributed to the less variation in data of class 2 defects. On the other hand, even though defect 3 has a larger database, the interclass variability results in a lower score.

Table 4.1 Accuracy scores and loss of multioutput model for each defect.

	Defect 1	Defect 2	Defect 3	Defect 4
Accuracy	0.9318	0.9804	0.8781	0.9535
Loss	0.2119	0.0717	0.3335	0.1474

5.4 Image segmentation

Figs. 4.14–4.16 illustrate the plots of dice coefficient and the validation loss as a function of epochs for the three trained segmentation models, that is, FCN model, Hyper-tuned Modified U-Net, and residual U-Net, respectively.

All the models were trained for 50 epochs with a batch size of 32. The baseline model, that is, the fully connected network gives a dice score of 0.5672. The loss value of the validation sets is like the training set for hyper-tuned modified U-Net as well as the residual U-Net, indicating that the models will generalize well with the other sets. Hyper-tuned modified U-

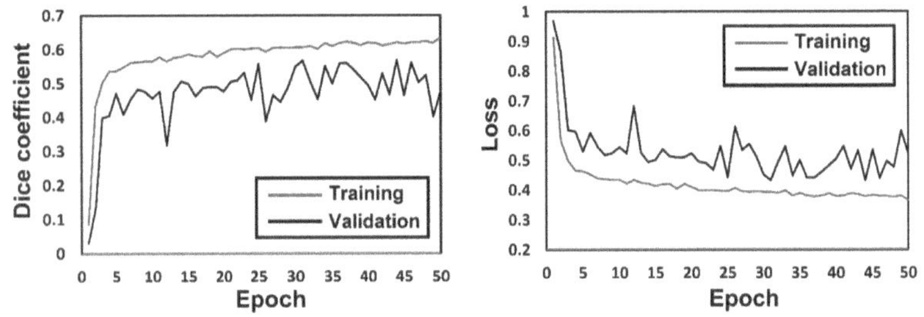

Figure 4.14 Dice and loss coefficient plots of the FCN model.

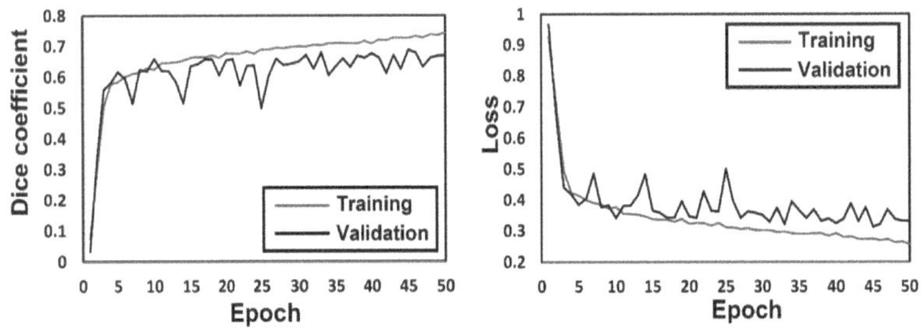

Figure 4.15 Dice and loss coefficient plots of the hyper-tuned modified U-Net model.

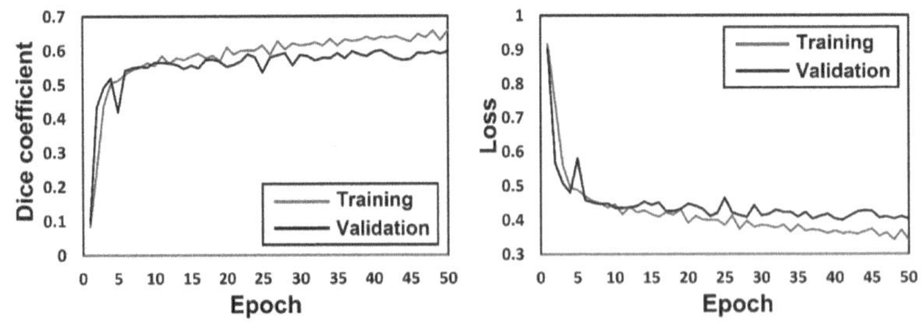

Figure 4.16 Dice and loss coefficient plots of residual U-Net model.

Table 4.2 Metrics of segmentation models.

S.No	Metrics	Proposed models		
		FCN	Hyper-tuned modified U-NET	Residual U-NET
1	Dice coefficient	0.5672	0.6888	0.60027
2	Precision	0.6065	0.7191	0.6941
3	Recall	0.5499	0.6682	0.5388
4	Mean IoU	0.4027	0.5277	0.4338

Net shows the best results with a dice coefficient of 0.69. The slow but a steady increase in the dice coefficient of residual U-Net indicates that training for more time may give better results. It is also important to note that the additional batch normalization and the dropout layers for hyper-tuned modified U-Net and the Residual U-net result in a much stable learning curve with minimal fluctuations in validation.

Table 4.2 summarizes the evaluation metrics used for comparing the segmentation models. Other than the dice coefficient, three other metrics, namely, precision, recall, and mean IoU scores are also monitored. Precision scores help in understanding as to how many relevant predictions the model can make. For example, whenever the hypertuned modified U-Net model predicts a pixel as being defective, it is correct 71.91% of the time. Recall scores provide an estimate of the model correctly identifying the pixels with defects. Mean IoU scores specify the amount of overlap between the predicted and the ground truth. Overall, the hypertuned modified U-Net model is superior in all aspects.

6. Conclusions

The surface imperfections of steel are the subject of this chapter's investigation. The dataset explored in the work is obtained from Severstal. The exploratory data analysis (EDA) of the dataset reveals that a single image may contain one or more defects, thus presenting a problem statement of multilabel classification and segmentation. EDA also demonstrates that there is a significant class imbalance within the dataset, which is addressed using data augmentation, class-wise reweighing and adaptive learning rate methods. A multioutput model is also trained to get an estimate of how the data imbalance is affecting the overall accuracy of a single output network. Since pixel-wise defect masks are also provided in the dataset, three multilabel segmentation models are trained with an aim of maximizing the dice coefficient. A fully connected network that achieves a dice score of 0.5672 is treated as the baseline model. The other two segmentation models are inspired by the well-reported U-Net architecture. The original encoder-decoder network with skip connections is hyper-tuned and modified by adding new learning layers and activation functions as part of the first modification. In the second method, the U-Net is strengthened with the help of residual learning. Among the segmentation models, the hyper-tuned modified U-Net design achieves a dice score of 0.69, whereas the Residual U-Net shows promising results as is observed from its smooth training curve. With a final dice score of 0.69 on the best model (which is an increase of 0.2 from the original U-Net structure), the objective of this study is successfully fulfilled.

The present investigation on defect detection of steel plates will be the foundation of our next research sojourn toward automated identification of defects and anomalies inside steel pipes of moderately smaller internal diameter (~50–60 mm). Preliminary conceptual designs for such mechatronic devices (in the form of miniature bio-robot) are under study at present. Complete validation of the developed methodology, based on deep learning technique, will be made with the working level firmware of the miniature bio-robot(s) in near future.

References

1. He K, Wang L. A review of energy use and energy-efficient technologies for the iron and steel industry. *Renew Sustain Energy Rev*. 2017;70:1022–1039. https://doi.org/10.1016/j.rser.2016.12.007.
2. Jiang Y, Ling T-C, Shi C, Pan S-Y. Characteristics of steel slags and their use in cement and concrete—a review. *Resour Conserv Recycl*. 2018;136:187–197. https://doi.org/10.1016/j.resconrec.2018.04.023.
3. Bulnes FG, Usamentiaga R, García DF, Molleda J. Vision-based sensor for early detection of periodical defects in web materials. *Sensors*. 2012;12(8). https://doi.org/10.3390/s120810788.
4. Yun JP, Choi SH, Kim SW. Vision-based defect detection of scale-covered steel billet surfaces. *Opt Eng*. 2009;48(3):1–9. https://doi.org/10.1117/1.3102066.
5. Martin D, Guinea DM, García-Alegre MC, Villanueva E, Guinea D. Multi-modal defect detection of residual oxide scale on a cold stainless steel strip. *Mach Vis Appl*. 2010;21(5):653–666. https://doi.org/10.1007/s00138-010-0260-5.
6. Takino H, Hosaka T. Shaping of steel mold surface of lens array by electrical discharge machining with single rod electrode. *Appl Opt*. 2014;53(33):8002–8005. https://doi.org/10.1364/AO.53.008002.
7. He Y, Song K, Meng Q, Yan Y. An end-to-end steel surface defect detection approach via fusing multiple hierarchical features. *IEEE Trans Instrum Meas*. n.d.;69(4):1493–1504. https://doi.org/10.1109/TIM.2019.2915404.

8. Jeon Y-J, Choi D, Lee SJ, Yun JP, Kim SW. Steel-surface defect detection using a switching-lighting scheme. *Appl Opt*. 2016;55(1):47–57. https://doi.org/10.1364/AO.55.000047.
9. Wang C, Yu Y, Yu J, Zhang Y, Zhao Y, Yuan Q. Microstructure evolution and corrosion behavior of dissimilar 304/430 stainless steel welded joints. *J Manuf Process*. 2020;50: 183–191. https://doi.org/10.1016/j.jmapro.2019.12.015.
10. Xi J, Shentu L, Hu J, Li M. Automated surface inspection for steel products using computer vision approach. *Appl Opt*. 2017;56(2):184–192. https://doi.org/10.1364/AO.56.000184.
11. Yoo H-J. Deep convolution neural networks in computer vision. *IEIE Trans Smart Process Comput*. 2015;4(1):35–43. http://www.dbpia.co.kr/journal/articleDetail?nodeId=NODE06217768.
12. Lu J, Tang S, Wang J, Zhu H, Wang Y. A review on object detection based on deep convolutional neural networks for autonomous driving. In: *Chinese Control and Decision Conference (CCDC)*. 2019:5301–5308. https://doi.org/10.1109/CCDC.2019.8832398.
13. Fu J, Zhu X, Li Y. *Recognition of Surface Defects on Steel Sheet Using Transfer Learning*. arXiv preprint arXiv:190903258; 2019.
14. Konovalenko I, Maruschak P, Brezinová J, Viňáš J, Brezina J. Steel surface defect classification using deep residual neural network. *Metals*. 2020;10(6). https://doi.org/10.3390/met10060846.
15. Damacharla P, Rao AMV, Ringenberg J, Javaid AY. TLU-Net: a deep learning approach for automatic steel surface defect detection. In: *International Conference on Applied Artificial Intelligence (ICAPAI)*. 2021:1–6. https://doi.org/10.1109/ICAPAI49758.2021.9462060.
16. Wang S, Xia X, Ye L, Yang B. Automatic detection and classification of steel surface defect using deep convolutional neural networks. *Metals*. 2021;11(3). https://doi.org/10.3390/met11030388.
17. Minaee S, Boykov YY, Porikli F, Plaza AJ, Kehtarnavaz N, Terzopoulos D. Image segmentation using deep learning: a survey. *IEEE Trans Pattern Anal Mach Intell*.n.d.:1-1. https://doi.org/10.1109/TPAMI.2021.3059968.
18. Takos G. *A Survey on Deep Learning Methods for Semantic Image Segmentation in Real-Time*. arXiv preprint arXiv:200912942; 2020.
19. Severstal: Steel Defect Detection; n.d. https://www.kaggle.com/c/severstal-steel-defect-detection/data. Accessed 2 July 2021.

CHAPTER FIVE

Extremum center interpolation-based EMD approach for fault detection of reciprocating compressor

Amitkumar Patil[1], Gunjan Soni[1], Anuj Prakash[2]
[1]Malaviya National Institute of Technology, Mechanical Engineering, Jaipur, Rajasthan, India
[2]Tata Consultancy Service, Bangalore, Karnataka, India

Contents

1. Introduction

With the advent of Industry 4.0 and associated technologies, condition-based maintenance strategy has attracted much attention. Especially, in high-risk work environments, such as petrochemical industry, off-site condition monitoring and fault diagnosis systems play critical role. The benefits of such condition-based health monitoring systems are twofold: First, personnel safety, which is priority in such environment, is achievable. Also, labor costs of onsite inspections can be minimized. Second, with advanced sensor technology and robust condition-based health monitoring and fault diagnosis systems, production-critical machinery can be monitored more effectively to ensure reliable plant operations. Reciprocating compressor is considered to be a

Smart Electrical and Mechanical Systems
ISBN: 978-0-323-90789-7
https://doi.org/10.1016/B978-0-323-90789-7.00012-9

critical component in petrochemical plant and processing industries, where it is commonly used for fluid compression. Due to its critical role in these industries, reliable operation of the compressor is a priority for maintenance engineers.[1] However, compressors work under highly dynamic operating conditions, which makes it quite difficult to monitor and diagnose faults.

In order to overcome these challenges, many condition monitoring and fault diagnosis approaches have been proposed. Majority of these reciprocating compressor studies are focused on valve fault diagnosis.[2] The primary reason for this inclination in researchers' interest is criticality and frequency of the fault occurrence. A study has provided that the majority (36%) of the unplanned shutdowns due to reciprocating compressors are caused due to valve failure.[2] As per the study,[3] primary causes for valve failure are dynamic working conditions, erroneous installation, and valve parameters' selection error. Considering the presented reasons, it is of significance to timely diagnosis of valve faults in reciprocating compressor before it severely damages the whole system. However, a major challenge in developing an accurate fault diagnosis model is signal nonstationarity. Due to external interferences and time-varying operating conditions, signals collected from the sensors mounted on the compressor are nonlinear and nonstationary in nature.[4,5] This makes the valve fault diagnosis a complex task.

To handle the nonstationarity of the signals, several signal processing techniques have been proposed in the literature compressor.[5] These techniques can be categorized on the basis of type of signals collected from different types of sensors, such as vibration signal analysis,[6] pressure signal analysisguerra,[7] temperature signal analysis,[8] acoustic signal analysis.[9] Time-frequency signal analysis techniques are known to effectively handle these nonstationary signals. Of these techniques, fast Fourier transform,[7] short-term Fourier transform (STFT),[10] wavelet transform[1,11] Wigner-Ville distribution (WVD) le[12] and EMD[13] are popular among researchers. While STFT provides limited resolution and WVD suffers from cross-term interferences, overcoming these limitations wavelet transform can effectively detect signal transients.[14] However, wavelet transform necessitates selection of optimal wavelet basis and number of layers for effective signal decomposition.[15] Due to adaptive signal decomposition abilities, EMD can produce higher time-frequency resolution compared to the other techniques.[14]

In existing studies, EMD has demonstrated promising results in detecting the impulses induced by localized faults in rotary components.[16,17] Another, EMD based fault diagnosis approach for clearance fault diagnosis of diesel engine.[18] EMD technique was successfully applied for gear root crack detection in Ref. 19. A modified EMD by replacing the interpolation method with B-spline fitting was used to gearbox fault diagnosis in Ref. 20. EMD has been successfully applied for reciprocating compressor fault detection in combination with autocorrelation function in Ref. 21. Later, EMD coherence-based fault diagnosis approach was proposed by the same authors in Ref. 13 for eliminating the cylinder signal interference. In,[22] rational Hermite interpolation-based EMD technique was applied to diagnose valve failure. Nevertheless, it is to be noted that EMD method is developed through a series of experiments and lacks a mathematical foundation. This method has a few issues, such as intrinsic mode aliasing, end effects, and overshooting and/or undershooting problems, that need to be studied further.[23] Trying to alleviate these limitations, several studies have attempted to develop improved interpolation methods for EMD process. In,[24] it was demonstrated that taut spline has advantages over the commonly used cubic spline. Later, it was established that cubic spline yields better results compared to the linear interpolation method[25] A piecewise power function was proposed by Qin and Zhong in Ref. 26. Piecewise parabola interpolation and piecewise linear fractal interpolation methods were also tested to improve the EMD procedure.[27,28] However, these methods could not improve the decomposition performance to great extension and still suffered through overshoot and undershoot phenomenon. Another attempt for attaining more flexibility in interpolation, a rational spline integrated with pole parameter was proposed in Ref. 29. In,[30] a convex constrained optimization-based envelope model was proposed that improved the EMD performance. A few attempts based on cubic Hermite interpolation were proposed in past[31,32] applied cubic Hermite interpolation for curve fitting integrated with double-iterative sifting process. The rational Hermite interpolation method, which is a cubic Hermite interpolation method, was proposed by adding control parameter modification in Ref. 31. However, a common drawback for the majority of these approaches is larger computational time, as it takes more time to interpolate upper and lower envelopes.

Huang J. et al.[33] addressed this challenge and developed an improved EMD procedure with Extremum Centered Interpolation (EMD-ECI).

The authors established that, for simulated as well as real signals, the EMD-ECI approach effectively tackles the mode mixing and overshoot-undershoot phenomenon. In spite of these advantages, this approach has not been tested in any of the existing reciprocating compressor fault detection studies. Hence, this study presents an improved fault diagnosis model, based on the EMD-ECI approach and statistical features.

The rest of the paper is structured as follows: First, theoretical background on EMD, EMD-ECI method, and statistical features chosen as system condition indicators is discussed. The following Section 3 presents the proposed fault diagnosis model. Next, experimental results and discussion are provided in Section 4. Section 5 provides the conclusion of the study.

2. Theoretical background

In this study, EMD-ECI technique is applied to evaluate its effectiveness in the signal decomposition compared to EMD. A brief theoretical background on both the techniques is provided as follows:

2.1 Empirical mode decomposition (EMD)

EMD is an adaptive signal decomposition technique that works on the basis of oscillatory components.[34] For identifying the fault information, EMD does not necessitates signal stationarity, periodicity or linearity information beforehand. EMD is used to decompose multicomponent signal into mono-component Intrinsic Mode Functions (IMFs).[35] IMFs are a set of frequency bands that constitute the multicomponent signal. As per the assumptions on the basis of which EMD is developed, any nonstationary signal can be expressed as sum of IMFs and residual component. It can be expressed in mathematical form as following:

$$x(t) = \sum_{n=1}^{N} \mathrm{IMF}_n(t) + \mathrm{res}(t) \tag{5.1}$$

where $\mathrm{IMF}_n(t)$ and res(t) denote nth IMF and residual component. It is to be noted that, to qualify as an IMF, a constituent function (extracted from raw signal) must satisfy the following two conditions: (1) the count of extrema and zero-crossings should be equal or defer by one at most, (2) mean value of envelope developed by maxima and minima must be zero at any point. The steps of EMD technique are as follows:

1. Detect all minima and maxima points of original signal $x(t)$.
2. Form lower envelope ($E_{lower}(t)$) and upper envelope ($E_{upper}(t)$) by using cubic spline interpolation technique for minima and maxima data points.
3. Calculate mean envelope by using the equation: $E_{mean}(t) = (E_{lower}(t) + E_{upper}(t))/2$.
4. Compute $s(t) = x(t) - E_{mean}(t)$.
5. Verify the two conditions for $s(t)$ if is qualified to be an IMF.
6. If $s(t)$ is qualified to be an IMF, then $IMF_n(t) = s(t)$ and repeat the procedure from step (1) for residual signal $res(t) = x(t) - s(t)$.
   Else, replace $x(t)$ with $s(t)$ and repeat the procedure.
1. Repeat steps (1–6) until monotonic res(t) is obtained.

The set of $IMF_n(t)$ contains frequency components such that $IMF_1(t)$ represents highest frequency component and subsequent IMFs express lower frequency components.

2.2 Empirical mode decomposition with extremum center interpolation (EMD-ECI)

In order to alleviate the overshoot and undershoot phenomenon in interpolation process, ECI method is used in this study. It was proposed by Huang J. et al.[33] in year 2015. This method finds all the extrema points of the signal, forms segments by connecting the minima and maxima points, finds the centers of the area covered by these minima-maxima envelopes, and fits a curve of these centers with cubic spline interpolation. The steps to execute this method are discussed briefly as follows:

1. Follow steps (1) and (2) from traditional EMD method as discussed in Section 2.1.
2. For each coordinate on horizontal axis ($x_i(t)$), corresponding vertical axis coordinates can be found on upper envelope and lower envelope which can be denoted by $y_{ui}(t)$ and $y_{li}(t)$. The center, named as extremum center, of the area covered by upper and lower envelope can be computed by Eq. (5.2):

$$y_{mi} = \frac{y_{ui} - y_{li}}{2} \tag{5.2}$$

   The y_{mi} are the extremum centers.

1. The extremum centers are fitted by cubic spline interpolation method. The fitted curve is called as extremum center interpolation curve, $m_1(t)$.
2. Calculate $s(t) = x(t) - m_1(t)$
3. Follow step (5–7) from traditional EMD method as discussed in Section 2.

2.3 Condition indicators

In order to extract fault information from the IMFs, various features have been used in previous study. Root mean square (RMS) and kurtosis are one of the most commonly used condition indicators (features) for compressor faults.[4] A short description on these condition indicators, used in this study, is presented below.

2.3.1 Root mean square (RMS)

RMS is the square root value of the mean of the sum of squares of signal. It is generally used to evaluate the signal amplitude and signal energy in time domain.[34] For signal $x(t)$, RMS can be computed by Eq. (5.3).

$$\mathrm{RMS} = \sqrt{\frac{1}{N}\sum_{t=1}^{N}(x(t))^2} \tag{5.3}$$

2.3.2 Kurtosis

Kurtosis is used for reflecting the impulse spikes present in a signal. It is fourth order normalized moment that can be calculated by Eq. (5.4).

$$k = = \frac{N^* \sum_{i=1}^{N}(x_i - x-)^4}{\left(\sum_{i=1}^{N}(x_i - x-)^2\right)^2} \tag{5.4}$$

Where, x_i denoted the ith datapoint in raw signal, $\overline{x}$ denotes mean of raw signal $x(t)$.

3. Proposed methodology

This section discusses the methodology applied in this paper. First, system signals are extracted with the help of respective suitable sensors. Next, the raw signals are decomposed into IMFs through the EMD-ECI method for further analysis. After signal decomposition, fault information-rich IMF components are selected from the obtained IMF-set. For monitoring the

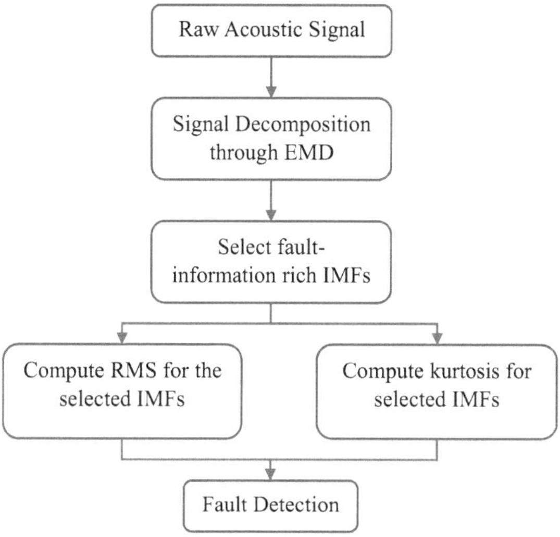

Figure 5.1 Proposed methodology.

system condition, statistical condition indicators (RMS and kurtosis) are extracted from the selected IMFs. Based on these features, fault diagnosis is performed. The following figure illustrates the steps of the proposed methodology (Fig. 5.1).

4. Experimental setup, results, and discussion

A single stage reciprocating compressor dataset used in Ref. 10 is used for validation of the EMD-ECI method for fault detection. The compressor had pressure range of 0–35 kg/cm^2 air pressure and was driven by 1440 rpm (415 V, 50 Hz, 5 A) induction motor. A Type PR-15 and pressure range of 100–213 psi pressure switch was used. The acoustic signal for healthy, inlet valve leakage (LIV) fault, and outlet valve leakage (LOV) fault states are collected from the experimental setup. A sample dataset of 50,000 datapoints was used in this study. The raw signal for each state (healthy, LIV, and LOV) is presented in Fig. 5.2.

The raw acoustic signals are decomposed with the help of EMD and EMD-ECI approaches first. It is observed that oscillatory modes (IMFs) reflecting the faults are present at higher power.[35] Hence, first IMF is selected for further analysis. For each state of the system, power spectrums

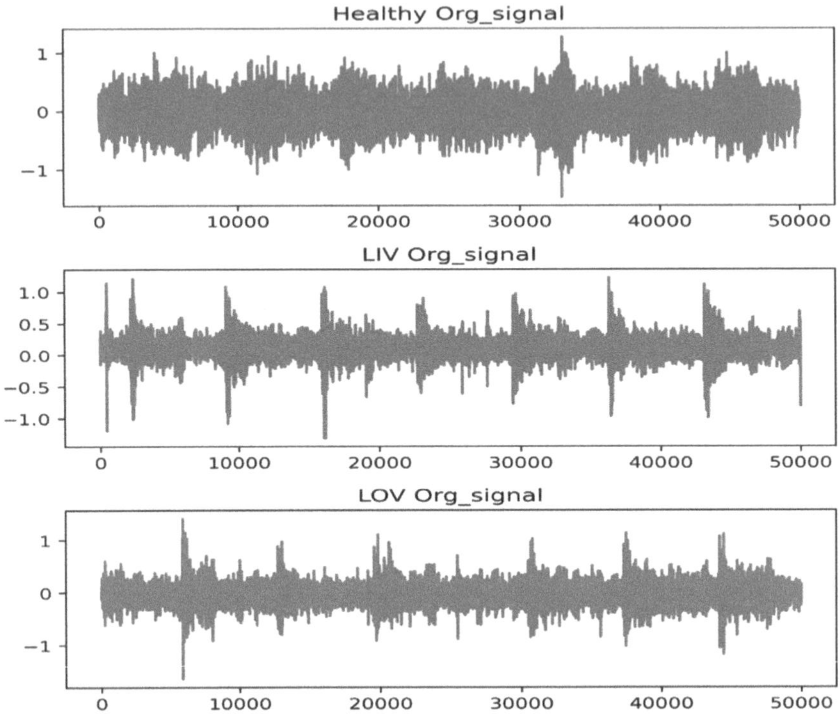

Figure 5.2 Raw acoustic signal of healthy, LIV, and LOV state of the compressor.

of IMF1 are computed and compared. From, and, it can be observed that EMD-ECI provides better denoising results compared to EMD. The sidebands of dominant frequency are clearly visible in EMD-ECI-based power spectrum compared to that of EMD based. For healthy and LIV fault state of compressor, the difference in dominant frequency is clearly noticeable in Figs. 5.3 and 5.4. The primary frequency of LIV fault state has more dominance in 1000 Hz, whereas healthy condition has denser frequency distribution in 500 Hz. The healthy and LOV fault can be differentiated based on the power of primary frequency.

However, it is not a convenient way fault detection. Therefore, two popular statistical condition indicators are employed in this study. RMS and kurtosis value for IMF1 to IMF6 obtained from EMD and EMD-ECI methods are illustrated in Figs. 5.5 and 5.6 respectively. From Fig. 5.6, it can be observed that, for both EMD and EMD-ECI extracted IMFs,

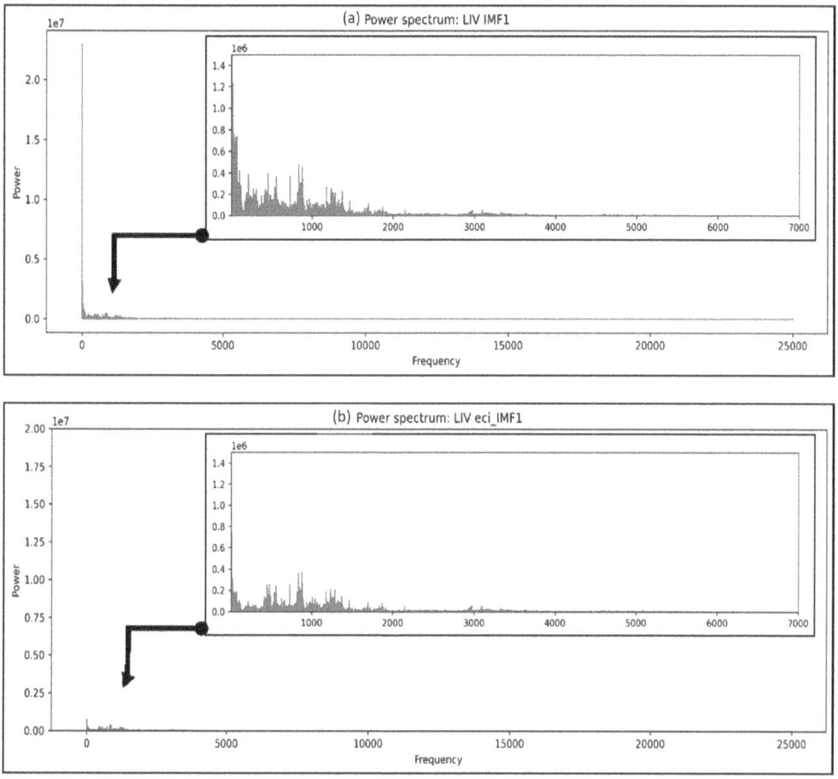

Figure 5.3 For LIV state: Power spectrum of IMF1 extracted with (A) EMD and (B) EMD-ECI.

RMS can differentiate the healthy state from faulty state. However, the latter approach shows more consistent results. Fig. 5.7 depicts that Kurtosis values for IMF1 to IMF6 extracted from EMD method, compared to EMD-ECI method, are not as consistent in differentiating the system conditions. It clearly shown that, EMD-ECI based kurtosis is comparatively more effective than the EMD-based kurtosis in separating the three system states. Moreover, it is also observed from the Fig. 5.7 that the kurtosis values of healthy state are lowest consistently, whereas that for LIV signals are highest. In case of EMD, the kurtosis values are evidently not reflecting any trend and overlapping in two cases. It can be stated from the results that EMD-ECI

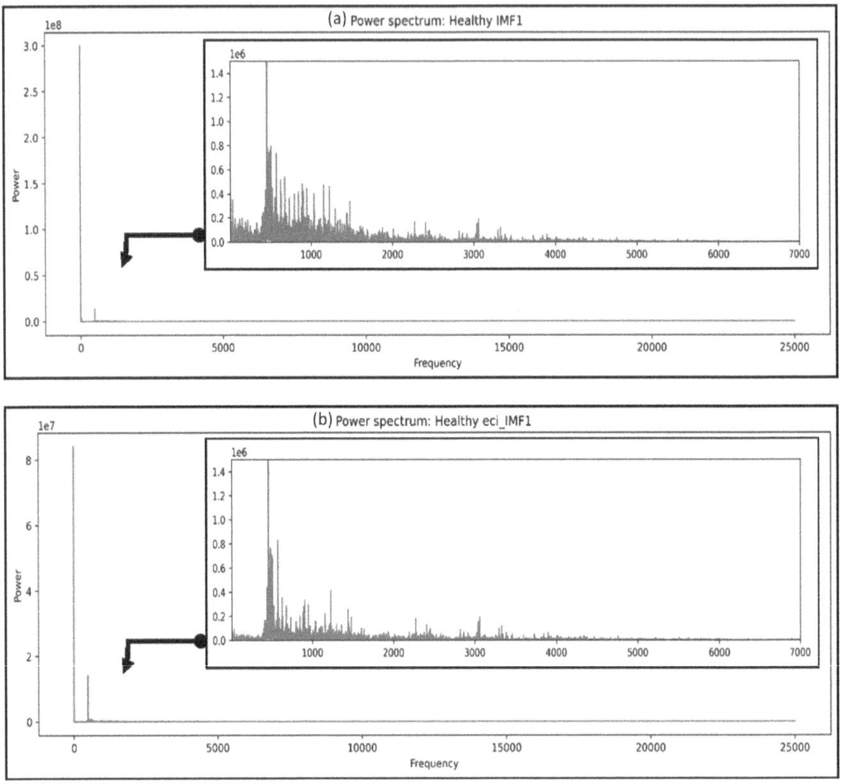

Figure 5.4 For Healthy state: Power spectrum of IMF1 extracted with (A) EMD and (B) EMD-ECI.

approach provides better signal decomposition and enhanced signal impulses (reflected by kurtosis) that further improves the fault detection ability.

5. Conclusion

In petrochemical industry, reciprocating compressor is a critical component as it works under highly dynamic operating conditions. Majority (around 36%) of all unplanned shutdowns in the plant are due to compressor valve failure. Hence, it is of significance to develop an accurate condition-based monitoring and fault diagnosis system for the compressor. A major challenge in developing in such a model is that the nonstationary nature of signals produced by compressor is difficult to analyze.

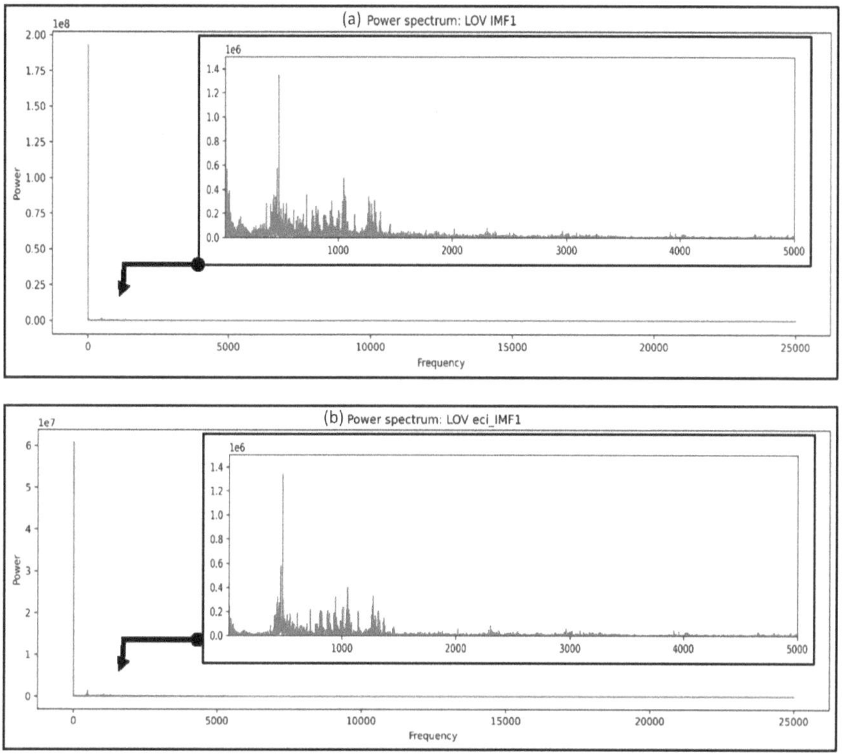

Figure 5.5 For LOV state: power spectrum of IMF1 extracted with (A) EMD and (B) EMD-ECI.

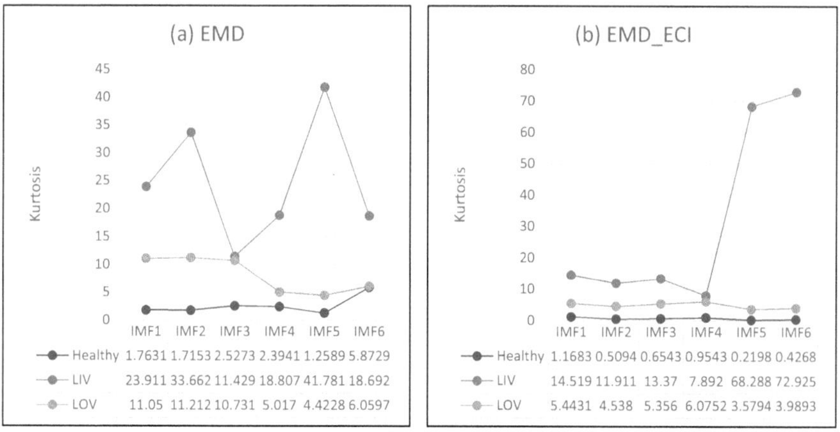

Figure 5.6 Kurtosis value of IMF1 to IMF6 extracted with (A) EMD and (B) EMD-ECI.

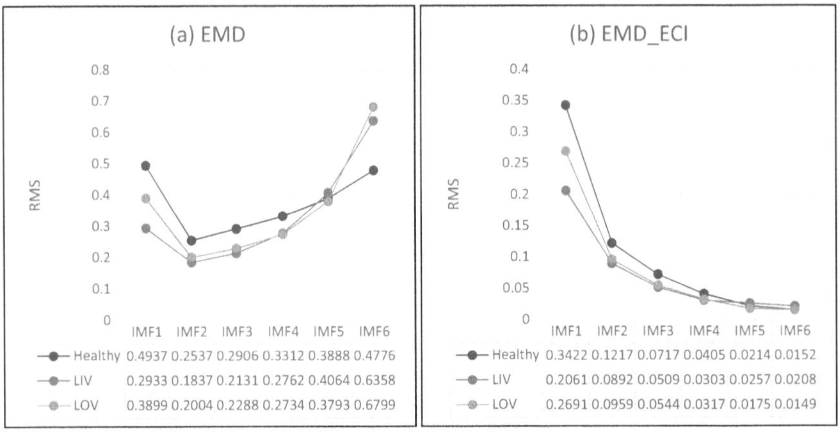

Figure 5.7 RMS value of IMF1 to IMF6 extracted with (A) EMD and (B) EMD-ECI.

EMD has been quite effective in analyzing these types of signals, however, suffers from overshoot and undershoot phenomena. To overcome this limitation, ECI method is employed in this study. For extracting the fault information from the processed signal, statistical features (RMS and kurtosis) are used. An acoustic sensor-based reciprocating compressor dataset collected for healthy, LIV fault, and LOV fault states is used to validate the proposed approach. The results show that EMD-ECI method can effectively denoise the signal as compared to EMD. RMS value obtained by EMD-ECI method has potential to distinguish the healthy state signals from fault signals. However, kurtosis value of EMD-ECI processed signal exhibits better results and cannot differentiate the LIV and LOV faults clearly. Observations from the results suggest that EMD-ECI method provides better signal decomposition results and further enhances the fault detection ability of the extracted features. However, the results may vary for different system operating under different working conditions. Hence, future studies can be focused on adopting the applied approach in data-driven remaining useful life prediction models and testing its effectiveness for other complex systems operating under dynamic environment.

References

1. Elhaj M, Gu F, Ball AD, Albarbar A, Al-Qattan M, Naid A. Numerical simulation and experimental study of a two-stage reciprocating compressor for condition monitoring. *Mech Syst Signal Process.* 2008;22(2):374–389. https://doi.org/10.1016/j.ymssp.2007.08.003.
2. Schirmer AG, Fernandes NF. On-line monitoring of reciprocating compressors. In: *NPRA Maintenance Conference*. 2004:1–20.

3. Motriuk RW. Reciprocating compressor valve failure - digital modelling and analysis. In: *Proceedings of the International Pipeline Conference, IPC*. vol. 2. 1996:993–1002. https://doi.org/10.1115/ipc1996-1907.
4. Sharma V, Parey A. Performance evaluation of decomposition methods to diagnose leakage in a reciprocating compressor under limited speed variation. *Mech Syst Signal Process*. 2019;125:275–287. https://doi.org/10.1016/j.ymssp.2018.07.029.
5. Zhao H, Wang J, Han H, Gao Y. A feature extraction method based on HLMD and MFE for bearing clearance fault of reciprocating compressor. *Measur J Internat Measur Confeder*. 2016;89:34–43. https://doi.org/10.1016/j.measurement.2016.03.076.
6. Pichler K, Lughofer E, Pichler M, Buchegger T, Klement EP, Huschenbett M. Fault detection in reciprocating compressor valves under varying load conditions. *Mech Syst Signal Process*. 2016;70–71:104–119. https://doi.org/10.1016/j.ymssp.2015.09.005.
7. Guerra CJ, Kolodziej JR. A data-driven approach for condition monitoring of reciprocating compressor valves. *J Eng Gas Turbines Power*. 2014;136(4). https://doi.org/10.1115/1.4025944.
8. Townsend J, Badar MA, Szekerces J. Updating temperature monitoring on reciprocating compressor connecting rods to improve reliability. *Eng Sci & Technol Internat J*. 2016;19(1):566–573. https://doi.org/10.1016/j.jestch.2015.09.012.
9. El-Ghamry MH, Reuben RL, Steel JA. The development of automated pattern recognition and statistical feature isolation techniques for the diagnosis of reciprocating machinery faults using acoustic emission. *Mech Syst Signal Process*. 2003;17(4):805–823. https://doi.org/10.1006/mssp.2002.1473.
10. Verma NK, Sevakula RK, Dixit S, Salour A. Intelligent condition based monitoring using acoustic signals for air compressors. *IEEE Trans Reliab*. n.d;65(1):291–309. https://doi.org/10.1109/TR.2015.2459684.
11. Peng ZK, Chu FL. Application of the wavelet transform in machine condition monitoring and fault diagnostics: a review with bibliography. *Mech Syst Signal Process*. 2004; 18(2):199–221. https://doi.org/10.1016/S0888-3270(03)00075-X.
12. Lee SK. Fault diagnosis of a gearbox using the sliced Wigner fourth order time frequency method smoothed by a new kernel function. *KSME Int J*. 1999;13(12): 940–947. https://doi.org/10.1007/BF03184761.
13. Wang L, Zhao JL, Wang FT, Ma XJ. Fault diagnosis of reciprocating compressor cylinder based on EMD coherence method. *J Harbin Inst Technol*. 2012;19(1):101–106.
14. Feng Z, Liang M, Chu F. Recent advances in time-frequency analysis methods for machinery fault diagnosis: a review with application examples. *Mech Syst Signal Process*. 2013;38(1):165–205. https://doi.org/10.1016/j.ymssp.2013.01.017.
15. Junsheng C, Dejie Y, Yu Y. Research on the intrinsic mode function (IMF) criterion in EMD method. *Mech Syst Signal Process*. 2006;20(4):817–824. https://doi.org/10.1016/j.ymssp.2005.09.011.
16. Cheng J, Yu D, Tang J, Yang Y. Application of frequency family separation method based upon EMD and local Hilbert energy spectrum method to gear fault diagnosis. *Mech Mach Theor*. 2008;43(6):712–723. https://doi.org/10.1016/j.mechmachtheory.2007.05.007.
17. Cheng J, Yu D, Tang J, Yang Y. Local rub-impact fault diagnosis of the rotor systems based on EMD. *Mech Mach Theor*. 2009;44(4):784–791. https://doi.org/10.1016/j.mechmachtheory.2008.04.006.
18. Li Y, Tse PW, Yang X, Yang J. EMD-based fault diagnosis for abnormal clearance between contacting components in a diesel engine. *Mech Syst Signal Process*. 2010;24(1): 193–210. https://doi.org/10.1016/j.ymssp.2009.06.012.
19. Loutridis SJ. Damage detection in gear systems using empirical mode decomposition. *Eng Struct*. 2004;26(12):1833–1841. https://doi.org/10.1016/j.engstruct.2004.07.007.

20. Liu B, Riemenschneider S, Xu Y. Gearbox fault diagnosis using empirical mode decomposition and Hilbert spectrum. *Mech Syst Signal Process*. 2006;20(3):718–734. https://doi.org/10.1016/j.ymssp.2005.02.003.
21. Wang L, Wang FT, Zhao JL, Ma XJ. Fault diagnosis of reciprocating compressors valve based on cyclostationary method. *J Donghua Univ*. 2011;28(4):349–352. http://www.ilib2.com/P-ISSN~1672-5220~2009~03.html.
22. Li Y, Xu M, Wei Y, Huang W. Diagnostics of reciprocating compressor fault based on a new envelope algorithm of empirical mode decomposition. *J Vibroengineer*. 2014;16(5): 2269–2286. http://www.jve.lt/Vibro/JVE.htm.
23. Guo T, Deng Z. An improved EMD method based on the multi-objective optimization and its application to fault feature extraction of rolling bearing. *Appl Acoust*. 2017; 127:46–62. https://doi.org/10.1016/j.apacoust.2017.05.018.
24. Huang NE, Shen Z, Long SR, et al. The empirical mode decomposition and the Hubert spectrum for nonlinear and non-stationary time series analysis. *Proc Math Phys Eng Sci*. 1998;454(1971):903–995. https://doi.org/10.1098/rspa.1998.0193.
25. Rilling G, Flandrin P, Goncalves P. On empirical mode decomposition and its algorithms. In: *IEEE-EURASIP Workshop Nonlinear Signal & Image Processing*. vol. 3. 2003:8–11.
26. Qin SR, Zhong YM. A new envelope algorithm of Hilbert-Huang Transform. *Mech Syst Signal Process*. 2006;20(8):1941–1952. https://doi.org/10.1016/j.ymssp.2005.07.002.
27. Fan Z, Hong T, Liu Z, Jing Z. Improve the envelope of EMD with piecewise linear fractal interpolation. *Key Eng Mater*. 2010;439–440:390–395. https://doi.org/10.4028/www.scientific.net/KEM.439-440.390.
28. Xu Z, Huang B, Li K. An alternative envelope approach for empirical mode decomposition. *Digit Signal Process A Rev J*. 2010;20(1):77–84. https://doi.org/10.1016/j.dsp.2009.06.009.
29. Pegram GGS, Peel MC, McMahon TA. Empirical mode decomposition using rational splines: an application to rainfall time series. *Proc Math Phys Eng Sci*. 2008;464(2094): 1483–1501. https://doi.org/10.1098/rspa.2007.0311.
30. Yang L, Yang L, Yang Z, Zhou F. A novel envelope model based on convex constrained optimization. *Digit Signal Process Rev J*. 2014;29(1):138–146. https://doi.org/10.1016/j.dsp.2014.02.017.
31. Li Y, Xu M, Wei Y, Huang W. An improvement EMD method based on the optimized rational Hermite interpolation approach and its application to gear fault diagnosis. *Measurement*. 2015;63:330–345. https://doi.org/10.1016/j.measurement.2014.12.021.
32. Jun-chang Z, Li Z. A Speech denoising method based on improved EMD. In: *International Conference on Multimedia and Signal Process*. vol. 2. 2011:305–309. https://doi.org/10.1109/CMSP.2011.149.
33. Huang J, Tang J, Zhang M, Zhang X, Han T. An improved EMD based on cubic spline interpolation of extremum centers. *J Vibroengineer*. 2015;17(5):2393–2409. http://www.jve.lt/Vibro/JVE.htm.
34. Sharma V, Parey A. A review of gear fault diagnosis using various condition indicators. *Procedia Eng*. 2016;144:253–263. https://doi.org/10.1016/j.proeng.2016.05.131.
35. Cho S, Shahriar MR, Chong U. Identification of significant intrinsic mode functions for the diagnosis of induction motor fault. *J Acoust Soc Am*. 2014;136(2):EL72–EL77. https://doi.org/10.1121/1.4885541.

CHAPTER SIX

Renewable energy sources forecasting and integration using machine learning

P.S.V. Kishore[1], Jami Rajesh[1], Sukanta Halder[2], Nakka Jayaram[1]
[1]Department of Electrical Engineering, National Institute of Technology Andhra Pradesh, Tadepalligudem, India
[2]Department of Electrical Engineering, Sardar Vallabhbhai National Institute of Technology, Surat, India

Contents

Abbreviations

Adj.R^2 adjusted correlation coefficient
ANFIS Adaptive Neuro-Fuzzy Inference System
ASH Ash content
BDTR Boosted Decision Tree Regression
BLR Bayesian Linear Regression
BP Back Propagation
CART Classification and Regression Tree

Smart Electrical and Mechanical Systems
ISBN: 978-0-323-90789-7
https://doi.org/10.1016/B978-0-323-90789-7.00005-1

CC Correlation coefficient
CNN Convolutional Neural Networks
DDPG Deep Deterministic Policy Gradient
DFR Decision Forest Regression
DMPC Diurnal Mean Power Curve
DTR Decision Tree Regression
FC Fixed Carbon
GBRT Gradient Boosted Regression Trees
GBT Gradient Boosting Tree
HHV Higher Heating Value
ILQR Iterative Linear Quadratic Regulator
kNN Kernal and Nearest Neighbor
KSI Kolmogorov Smirnov Integral
LASSO Least Absolute Shrinkage Selector Operator
LR Linear Regression
MAE Mean Average Error
MAE Mean Absolute Error
MARPE Maximum Absolute Relative Percentage Error
MASE Mean Absolute Scaled Error
MBE Mean Bias Error
MedAE Median Absolute Error
MHE Mean Huber error
MLP Multilayer Perceptron
MLR Multiple Linear Regression
MRE Mean Relative Error
MSE Mean Square Error
NNR Neural Network Regression
NRSE Normalized Root Square Error
NS Nash Sutcliffe Efficiency Coefficient
NWP Numerical Weather Prediction
oLSTM LSTM-Dropout
OT Operating Time
per20 Proportion of relative error <20%
PILCO Probabilistic Inference for Learn Control
PLS Partial Least Squares regression
PPO Proximal Policy Optimization
PR Polynomial Regression
r Correlative coefficient
R^2 Coefficient of determination
RAE Relative Absolute Error
RNMSE Rooted Normalized Mean Square Error
rRMSE Relative Root Mean Squared Error
SAE Stacked Auto Encoders
SARSA State Action Reward State Action
SMGRT Simple Membership Function and Fuzzy-
SVMR Support Vector Machine Regression
SVR Support Vector Regression
U1 Theil U-statistic one

U2 Theil U-statistic two
XGBoost eXtreme Gradient Boost

1. Introduction

Machine learning (ML) trains computers to think like human beings. Algorithms used in ML learn the information from previous input/output data without relying on equations, identifying patterns, and making decisions. After these algorithms are trained, it will predict the output values for a new input value. The performance of these algorithms will increase with the increase in the trained data.[1] There are so many techniques available to train computers. All these techniques make use of previously available data and accomplish different jobs. These techniques are mainly classified into three categories. They are

Supervised Learning: It is one of the ML techniques in which machines are trained using previous input-output data. Based on this data, the ML algorithm finds the relationship between inputs and outputs. Now this machine is ready to predict the output for new input.

Unsupervised Learning: It is a subset of ML in which raw data without any labels is given to the machine. The ML algorithm categorizes this unlabeled raw data into groups based on some similarities. Once it is trained, it will predict the group of the new input entry.

Reinforcement Learning: It is the third type of ML where an agent communicates with the environment and modifies its actions in order to maximize the cumulative reward. It does not require labeled or unlabeled data. The classification of different ML algorithms is shown in Fig. 6.1.

The applications ML can be in many fields, such as selfdriving cars, product recommendations, and fraud detections.[2] The data preprocessing methods are very important in developing ML models so as to increase the efficiency of the prediction. Data preprocessing is the process of making raw data in a systematic manner suitable for ML models.

The fast growth of industrial development throughout the world will intensify the decrease in reserves of fossil fuels and have a detrimental effect on the climate. Rising health hazards and threats from global climate change will result from these impacts. Renewable energy technology is growing very fast at the moment. Renewable energy means energy that is not depleted when used. Different types of renewable energies are present, such as solar, wind, hydro, biomass, tidal, and geothermal, etc. Plenty of

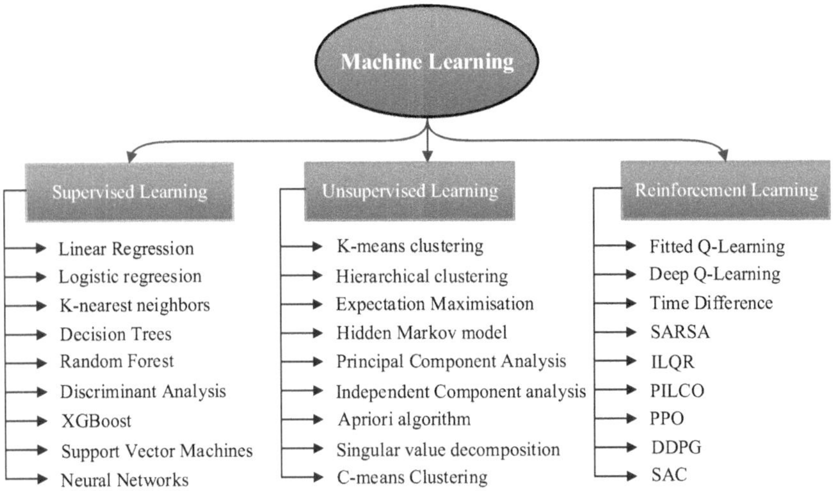

Figure 6.1 Machine learning algorithms classification.

studies have been done on renewable energies due to their low environmental pollution and sustainability. The implementation of renewable energy systems would be able to resolve critical problems facing current energy challenges, such as enhancing energy supply reliability and resolving regional energy shortages.

Most of the renewable energies are intermittent in nature. For example, solar energy is present only in the daytime and wind is present mostly in the nighttime and in hilly areas. Thus, the output of renewable energy systems depends on environmental circumstances. As the usage of renewable energy source (RES) is growing, it is important to develop suitable technologies to use these renewable energies effectively. By considering the variable load in the system and the environmental dependence of RES, it is very important to forecast the RESs for the sustainable development of renewable energy systems.[3] Various studies reveal that ML models will help us in the forecasting of RESs. Accuracy of prediction of RESs will be increased with hybrid ML models. Different periods were used to forecast renewable energy, such as minutes, hours, days, and weeks, according to different purposes. Precision and forecasting efficiency has generally been used to test ML models' success in renewable energy forecasts.[4] Quantile-based probabilistic forecasting and other forecasting models were developed for renewable energy generation in Ref. 5.

Owing to the unpredictable existence of all of the RESs, it is a major challenge today to be able to integrate greater proportions of renewable power into the energy mix. As renewable energy penetration has increased, some of the key technological potential challenges that have been observed are load management, power system transients, and harmonics, reactive power, power system security, low power factor, voltage fluctuation, storage system, forecasting, and scheduling, and electromagnetic interference.[6] ML is used in many parts of the power system. Fig. 6.2 shows the use of ML in different divisions of power systems.

This chapter summarizes the different ML techniques used to predict the generation of RESs. This chapter also reviews the different ML algorithms used for the integration of RES into the grid, such as operations and management of the grid. This chapter also focuses on how the MPP point is maintained using several ML techniques.

The rest of the chapter is as follows. Section 2 describes the various ML techniques used in the prediction of solar power, wind power, and other renewable sources. Section 3 explains the various ML techniques used in determining plant location, size, and configuration. Management of grid using different ML techniques is presented in Section 4. Section 5 describes how the maximum power can be extracted from renewable sources.

2. Prediction of renewable energy sources

2.1 Prediction of solar power

Solar power is highly dependent on atmospheric conditions such as irradiation, temperature, moisture in the air, and wind speed. These

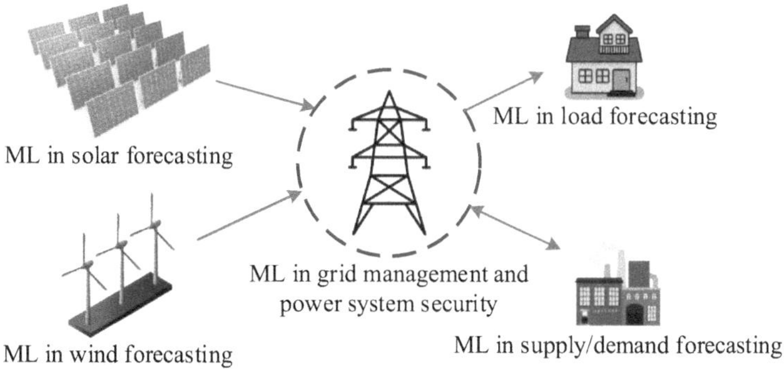

Figure 6.2 Use of machine learning in power system.

atmospheric conditions are not the same throughout the day as well as throughout the year. This affects the stability and reliability of the solar power generation system. Therefore, forecasting solar power in the near future is essential for the smooth operation of solar-powered systems. There is a lot of research happening from the last decade in the forecasting of solar power. Various ML models were developed to accurately forecast solar power in the near future.

One hour step ahead solar radiation is forecasted by a new method called SVM with time series principle.[7] Levenberg–Marquardt backpropagation algorithm-based deep learning model[8] is used to train and test the ANN models to predict the solar irradiance hourly. The power output of a PV plant is forecasted for the next 2–4 h using the ANN model.[9] A ML model of RNN and deep learning models of LSTM and GRU are used in Ref. 10 to predict the very short-term solar irradiation. Day-ahead prediction of solar power was done in Ref. 11–14. A new method based on ANN merges the numerical weather prediction methods and statistical learning methods in Ref. 11 to forecast the production of power from PV plants in Argentina. Multiple regression techniques like linear least squares and SVM using multiple kernel functions[12] are used in developing prediction models. Dense convolutional network and residual network are the two deep CNNs, which are proposed in Ref. 13 as forecasting models. In Ref. 14, four ML algorithms named SVM, ANN, kernel and nearest neighbor (k-NN), and deep learning are used to predict global solar radiation. Two ML techniques, namely, simple linear regression and M5P regression tree are compared in Ref. 15 for the prediction of solar power generation. A new solar irradiance forecasting model is proposed based on an extreme learning machine (ELM) in Ref. 16 using the historical irradiance data. In Ref. 17, four ML algorithms were developed and compared to predict the PV power generation. Different models were developed based on DL, SMGRT, ANFIS by considering different parameters in Ref. 18 for the estimation of solar power generation. A new AI algorithm, namely, RE-SOINN was proposed in Ref. 19 for accurate and adaptive solar irradiance forecasting even for highly fluctuating profiles (when there are fast-moving clouds). Probabilistic forecasting of solar generation using a direct deep learning architecture is proposed in Ref. 20. The thermal power of a solar power tower system is forecasted half an hour ahead using a hybrid method that combines mechanism modeling and deep learning.[21] Six different ML models were developed and compared in Ref. 22 for forecasting solar radiation. An advanced case-based learning and k-means clustering are the two ML models used to estimate the monthly average daily solar radiation.[23] In Ref. 24, the authors intend to make prediction models for solar irradiance LR, RT, SVM models which use previous weather data such as wind speed, temperature, humidity, and atmospheric pressure. A new technique named Quantile Regression Averaging is used for forecasting solar power in a short interval of time.[25,26] A few articles related to the prediction of solar power using ML algorithms are shown in Table 6.1.

2.2 Prediction of wind power

Wind power is one of the most promising RESs as it is clean, safe, and inexhaustible. Over the past decade, wind power has been growing as an intermittent RES, and the integration of wind power into the grid has very adverse effects on the reliability, protection, and power quality of the power grid. Metrological factors such as wind speed, air density, and precipitation affect the power generation from wind energy systems. For proper planning and scheduling of the load for the grid, wind power forecasting is important. This gives us a great advantage in ensuring successful grid management, thus reducing losses and thus reducing the cost of power generation. ML algorithms help in forecasting wind power accurately.

The time scale for wind can be further categorized as follows. If the forecasting time is less than 1 hour, it is considered as ultrashort-term forecasting; if the forecasting time is about 14–15 h, it is named as short-term forecasting; if the time frame is about 1 week, then it is termed as medium-term forecasting; if the time of forecasting is more than a year, then it is considered as long-term forecasting. An Ensemble ML technique is used to develop 1 hour ahead wind speed forecasting model.[27] A wind power forecasting model was developed in Ref. 28 using a random forest algorithm by taking spatially averaged wind speed and wind direction as inputs. Day-ahead forecasting of wind speed is done using the convolutional neural networks in Ref. 29. In Ref. 30–34, different ML models such as RF, SVM, DT, BP, RVM, LR, SVR, ELM, and LS-SVM are used to predict wind power in the short term. There are different types of parameters that can be chosen as the input features of the ML models. A short-term wind speed forecasting model using LSSVM was developed in Ref. 35. The input parameters of this model are optimized using an ant colony algorithm. Several deep learning architectures were developed in Ref. 36 based on MLP, CNN, and RNN to forecast wind speeds at a 12-hour horizon. Day-ahead wind power was predicted in Ref. 37 by using a novel EDCNN

Table 6.1 Literature on prediction of solar power using ML.

Forecasting time	Paper and year	ML technique	Parameters	Performance indicator
Hourly	Belaid et al. 2020[7]	SVM with time series principle	Relative humidity Rh(%), temperature T(°C), pressure P(Pa), wind speed V(m/s), and hourly global solar radiation(HGSR) (Wh/m^2)	KSI, RMSE, NRMSE, NMBE, MAPE, and R
	Bamisile et al. 2020[8]	DL	Year, month, day, hour, solar irradiance, sun elevation, ambient temperature, and wind speed	R, R^2, RMSE, and MAE
2–4 h	Yildiz and Acikgoz 2021[9]	Kernel extreme learning machine	Temperature, humidity, solar radiation, wind speed, pressure, cloud cover, and precipitation.	DMPC, R, RMSE, and MAE
Very short term	Bendali et al. 2020[10]	Recurrent neuron network, LSTM, and GRU	Solar irradiation	MSE, RMSE, and MAE
Day	Parra Raffán et al. 2019[11]	ANN	Maximum temperature, wind direction, wind speed, humidity, atmospheric pressure	RMSE
	Hassan et al. 2018[12]	Linear least squares and SVM using multiple kernels	Sunshine and radiation	MAE, RMSE

	Zang et al. 2020[13]	Deep convolutional neural networks	Sunny, partially cloudy, rainy, and overall	MAE, MSE, MSLE, NIA, MASE U1, U2 and MHE
	Ağbulut et al. 2021[14]	SVM, ANN, k-NN, and DL	Daily extraterrestrial solar radiation, cloud cover, solar radiation, daily minimum and maximum ambient temperature, and day length.	MAPE, RMSE, rRMSE, R^2, MBE, MABE, and t-statistics.
NA	Benhmed et al. 2018[15]	SLR, and M5P regression tree	Irradiance, dust, temperature, PV temperature, wind speed, and relative humidity	MAE, R, RMSE, RAE %, RRSE%
	Burianek and Misak 2016[16]	ELM	Clouds low, clouds high, clouds medium, clouds total, wind properties, humidity, atmospheric pressure, and temperatures.	RMSE, RNMSE
	Ağbulut et al. 2020[17]	SVM, ANN, kNN, and DL	Clear sky index solar irradiance, solar output power, and cloud position	R^2, RMSE, MBE, t-statistics, and MABE
	Üstün et al. 2020[18]	DL	Air temperature, sunshine duration, extraterrestrial solar radiation, cloudiness, soil temperature, and relative humidity	MBE, MSE, RMSE, and R^2

(Continued)

Table 6.1 Literature on prediction of solar power using ML.—cont'd

Forecasting time	Paper and year	ML technique	Parameters	Performance indicator
	Puah et al. 2021[19]	RE-SOINN	Power, denoising control, maximum age, lambda, and number of neighbors	MASE, RMSE
	Afrasiabi et al. 2020[20]	DL	Temperature, humidity, wind speed, PV power generation, and solar irradiance	MAPE and RMSE
	Wang et al. 2020[21]	DL	Solar azimuth angle, solar altitude angle, relative humidity, barometric pressure, and wind speed	RMSE and MAE
	Alizamir et al. 2020[22]	GBT, ANFIS-SC, ANFIS-FCM, MARS, ANN, CART	Minimum air temperature, maximum air temperature, wind speed, and relative humidity	RMSE, R, MAE, and NS
	Koo et al. 2019[23]	k-means clustering	Monthly average daily solar radiation	MAPE
	Javed et al. 2019[24]	LR, regression trees, and SVM	Atmospheric pressure, wind speed, humidity, and temperature	R^2

algorithm. Forecasting of short-term wind power is done in Ref. 38 using a combined SAE-BP algorithm, and the results are compared with BP and SVM methods. The proposed SAE-BP method gives the minimum forecasting error with high accuracy. An algorithm based on CNN is proposed in Ref. 39 to forecast the velocity of wind for the next 3 days by taking the past 1 week's wind speed data. A model involving a combination of RNN regularization methods using dropout and LSTM named oLSTM is presented in Ref. 40 to predict short-term wind power. A novel model based on long-short term memory was developed in Ref. 41 for the effective forecasting of wind power. This model performs better in comparison with the conventional ML models such as LR, RF, and GB. An improved back propagation neural network algorithm was proposed in Ref. 42 for wind power forecasting. In Ref. 43 annual wind energy density was estimated using a multilayer regression model, these results are found to be more accurate when compared with the PLS and SVR algorithms. Three ML models such as ANN, SVR, and GPR are compared in Ref. 44 for predicting solar power, wind power, and load forecasting. In Ref. 45 two one-dimensional CNNs were proposed for predicting the dominant wind speed and direction by utilizing historical data. In Ref. 46, authors used a convolutional network that uses a special feature named spatiotemporal feature for the prediction of wind power. The daily standard deviation of hourly wind speed, daily wind speed generated, and daily wind power data[47] are chosen as inputs to the ML models for predicting long-term wind speed by using LASSO, kNN, XGBoost, RF, SVR models. Forecasting of the speed of the wind using Quantile Regression was done in Ref. 48. Table 6.2 gives the various ML models, different parameters used for these models as inputs, and the performance indicators used in the literature are shown.

2.3 Prediction of other RES

2.3.1 Hydro

The water level in the reservoir is essential for sustainable hydropower generation. Accurately forecasting the level of water in the tank was done in Ref. 49 using different regression techniques named decision forest, boosted decision tree, neural network, and Bayesian linear. An improved LSSVM model is employed in Ref. 50 to forecast the level of water daily in the Yangtze River, China. Four models such as ANFIS, SVM, RBNN, and GRNN were developed and compared in Ref. 51 to estimate the level of the dam reservoir. The authors showed that the ANFIS model gave the best results among all four models. Four-hour-ahead forecasting of the water level in the reservoir is done in Ref. 52 by taking previous 2-hour inputs using ANN, SVM, and ANFIS models.

2.3.2 Tidal

Tidal energy is also a promising source of renewable energy that can be produced by energy obtained from the tides in the ocean. In Ref. 53, SVR is proposed to predict the direction and speed of the tidal current with high accuracy and is compared with the conventional method based on wavelet transform.

2.3.3 Geothermal

Geothermal energy is heat within the earth. As heat is continuously generated within the earth, geothermal energy is a source of renewable energy. The prediction of geothermal energy production using the LSTM encoder-decoder model is proposed in Ref. 54. In this, the LSTM encoder and LSTM decoder are used to learn the past production and to predict the future production of geothermal energy, respectively. The theoretical very Shallow Geothermal Potential (vSGP) is estimated in Ref. 55 using random forests by extracting the features of thermal conductivity, thermal diffusivity, and temperature at a particular location.

Very few articles are present in the literature on predicting hydro, tidal, and geothermal energies using ML models, which are given in Table 6.3.

2.3.4 Biomass

Biomass energy is also a nonconventional source of energy that is produced by plants such as maize and soy. Because of its capacity to use waste materials, biomass gasification is a viable power-generating method. Four regression techniques, namely PR, SVR, DTR, and MLP are employed in predicting biogases like CO, CO_2, CH_4, H_2, and HHV.[56] The prediction of HHV of biomass materials is done in Ref. 57 using the GBRT algorithm, which is trained using the biomass samples. The H_2 concentration in biomass gasification is predicted in Ref. 58 using the different supervised ML algorithms such as LR, KNN, SVMR, DTR. Articles in the literature about the prediction of biomass gases using ML are given in Table 6.4.

Table 6.2 Literature on prediction of wind power using ML.

Forecasting time	Paper and year	ML technique	Parameters	Performance indicator
One hour	Feng et al. 2017[27]	SVR, ANN, GBM, RF	Wind speed, temperature, humidity, and pressure	NMAE, NRMSE
	Lahouar and Ben Hadj Slama 2017[28]	RF	Spatially averaged wind speed and wind direction	MAE, RMSE, MXE, NMAE, MAPE, MASE, CC
Day	Hong and Satriani 2020[29]	CNN	Wind speeds	RMSE, NMSE, MAPE, R^2
Short term	Chaudhary et al. 2020[30]	RF, SVM, DT	Wind speed, air density, temperature, and precipitation	RMSE and MAPE
	Zhang et al. 2020[31]	Combined model of BP, wavelet and RVM model	Wind speed	RMSE, MAE, and OT
	Ali et al. 2018[32]	LR and SVR	Wind direction, wind speed, electrical power, turbine working state, and temperature	MAE, RMSE
	Mahmoud et al. 2018[33]	ELM	Wind speed, air density and air temperature	RMSE, MAE
	Zhang et al. 2017[34]	LS-SVM	Wind speed direction	MAE, RMSE, MAPE
	Li et al. 2019[35]	LSSVM	–	MSE, MRE
	Manero et al. 2019[36]	MLP, CNN, RNN	Wind speed, temperature, humidity, pressure, and direction	R^2
	Mujeeb et al. 2020[37]	EDCNN		MAE, NRMSE and MAPE

(Continued)

Table 6.2 Literature on prediction of wind power using ML.—cont'd

Forecasting time	Paper and year	ML technique	Parameters	Performance indicator
			NWP, wind speed, dry bulb temperature, dew point temperature	
	Jiao et al. 2018[38]	SAE-BP	—	MAE, RMSE, MAPE
	Huang and Kuo 2018[39]	CNN	Wind speed	MAE, RMSE
	Eze et al. 2019[40]	oLSTM	Wind speed	MSE, RMSE
NA	Donget al. 2019[41]	RNN structure model with LSTM units	Wind speed	MSE
	Wang et al. 2017[42]	BP neural network	Wind speed, wind direction, temperature, pressure	RMSE, MAE, r, per20
	Aksoy et al. 2019[43]	PLS, SVR, and MLR	Temperature, wind speed, and direction	R^2, MAE, MSE, and MedAE
	Sharifzadeh et al. 2019[44]	ANN, SVR, GPR	Seasonal variable, hourly variable, wind speed, wind power, and temperature	NRSE
	Harbola and Coors 2019[45]	1D CNN	Wind speed and direction	SMAPE
	2019[46]	CNN	–	MSE and RMSE
Long term	Demolli et al. 2019[47]	LASSO, kNN, xGBoost, RF, and SVR	Generated daily wind power data, daily standard deviation of hourly wind speed, and daily wind speed.	R^2, RMSE, and MAE

3. Determination of plant size, location, and configuration

Renewable energy power plants, unlike coal-fired plants, natural gas, and diesel need a wide area to run. A traditional power plant with an equivalent size takes up just a few square meters. Therefore, it is important to determine the plant size needed in relation to the necessity of energy. Solar power plants require a special orientation and location, as solar panels must be exposed to solar radiation to consume the maximum amount of electricity. The traditional PV system sizing methods (numerical, analytical, and empirical methods) can estimate the area required by the PV arrays and storage capacity for a given location, but they require several parameters such as altitude, longitude, daily solar irradiation, load, standalone PV system characteristics, and panel inclination. These methods require a significant amount of computation time to estimate the optimum combination. Therefore, ML methods can estimate storage capacity and PV array area with minimum inputs. Different artificial intelligence techniques used in the sizing of PV systems are discussed in Ref. 59.

Authors in Ref. 60 proposed an optimal sizing using a multiobjective function for a hybrid system containing wind and solar. A trained ANN is used in Ref. 61 to predict the sizing parameters of PV systems such as storage systems and PV array area with less number of inputs (the monthly average irradiation, the longitude, and the altitude). Authors claim that a forecast with a maximum error of 6% was made. A model combining MLP and infinite impulse filter was developed in Ref. 62 and it was trained using the data corresponding to 200 locations. This model estimates the optimal sizing combinations which allow the users to determine the storage batteries and the number of PV panels required for the given consumption. The study[63] uses a genetic algorithm to optimize the configuration of power-producing systems such as batteries, wind turbine generators, diesel generators, and PV systems. The authors claim that, as compared to diesel generators, the operating costs may be lowered by around 10%. In Ref. 64, the authors developed a strategy for optimum distribution and size of solar grid-connected systems in feeders using a multiobjective optimization approach, which provides the best compromise for both

Table 6.3 Literature on prediction of other RES using ML.

Type of RES	Paper and year	ML technique	Performance indicator
Hydro	Sapitang et al. 2020[49]	BDTR, DFR, BLR and NNR	R^2, MAE, RMSE, RAE, RSE
	Guo et al. 2019[50]	Improved LSSVM	RMSE, MAPE, d
	Üneş et al. 2019[51]	ANFIS, SVM, RBNN, GRNN	MSE, MAE, R
	Zhang et al. 2016[52]	ANN, SVM, ANFIS	NC, RMSE, R^2
Tidal	Kavousi-Fard and Su 2017[53]	SVR	MAPE, MARPE, RMSE
Geothermal	Gangwani et al. 2020[54]	LSTM encoder-decoder	—
	Assouline et al. 2019[55]	RF	RMSE, NRMSE

Table 6.4 Literature on prediction of biogases using ML.

Paper and year	ML technique	To predict	Parameters	Performance indicator
Elmaz et al. 2020[56]	PR, SVR, DTR, MLP	CO, CO_2, CH_4, H_2 and HHV	–	k-fold cross- validation, R^2, Adj.R^2, RMSE and NRMSE
Samadi et al. 2019[57]	GBRT	HHV of biomass	VM, FC, and ASH	MAPE, RMSE, R^2
Ozbas et al. 2019[58]	LR, KNN, SVMR, DTR	H_2 concentration	CO, CO_2, time, CH_4, heating value, O_2, and temperature	R^2, MSE, MAE, variance

technological and economic factors. In Ref. 65 proposes a flexible fuzzy data envelopment analysis-ANN technique for optimizing the placement of solar plants with probable uncertainty and complexity. In Ref. 66, the authors calculate the optimum size of the hybrid system (PV/wind/battery/diesel generator) using the Clonal Selection Algorithm based on the neural forecasting model. A hybrid optimization model was developed in Ref. 67 for optimal sizing and investment of a vanadium redox flow battery. For optimized sizing and allocation of distributed generation-based renewable sources for radial distribution systems, an ant lion optimization algorithm is proposed in Ref. 68. A radial distribution system consisting of 69 buses is used to test the proposed algorithm. According to the authors, their suggested algorithm is compared to others to emphasize its merits in decreasing overall power losses and, as a result, optimizing net savings.

The optimal size of the battery is calculated in Ref. 69 using an optimization problem with a nonlinear objective function and two basic constraints. This approach combines the primary aging mechanisms observed in actual batteries. The optimal design of a biomass/battery/biogas/solar-based hybrid system was done using the Gray Wolf optimization method[70] to provide continuous electricity to different households in some villages in Haryana, India. The proposed model's findings were considered to be superior to those obtained from harmony search and particle swarm optimization. In Ref. 71, by taking into account varied profiles of loads and PV, a rapid and accurate energy loss assessment approach in distribution systems utilizing the regression tree technique was presented. The suggested approach's efficacy has been shown on the 33-bus system utilizing multiple scenarios (clear day, cloudy day, and annual) with varied time resolutions. The suggested method may be used to analyze the long-term effects of PV, as a result, appropriately scale PV to reduce energy losses in distribution networks.[72] uses an approach based on the application of a sequential random Monte Carlo simulation to the system model to size a stand-alone PV system. The simulation findings aid in the sizing of solar panels and batteries based on the needed degree of dependability, while taking into consideration uncertainties such as energy consumption, solar radiation, and component failures. The basic flow chart of optimizing, sizing, and configuration of a PV plant is shown in Fig. 6.3.[73]

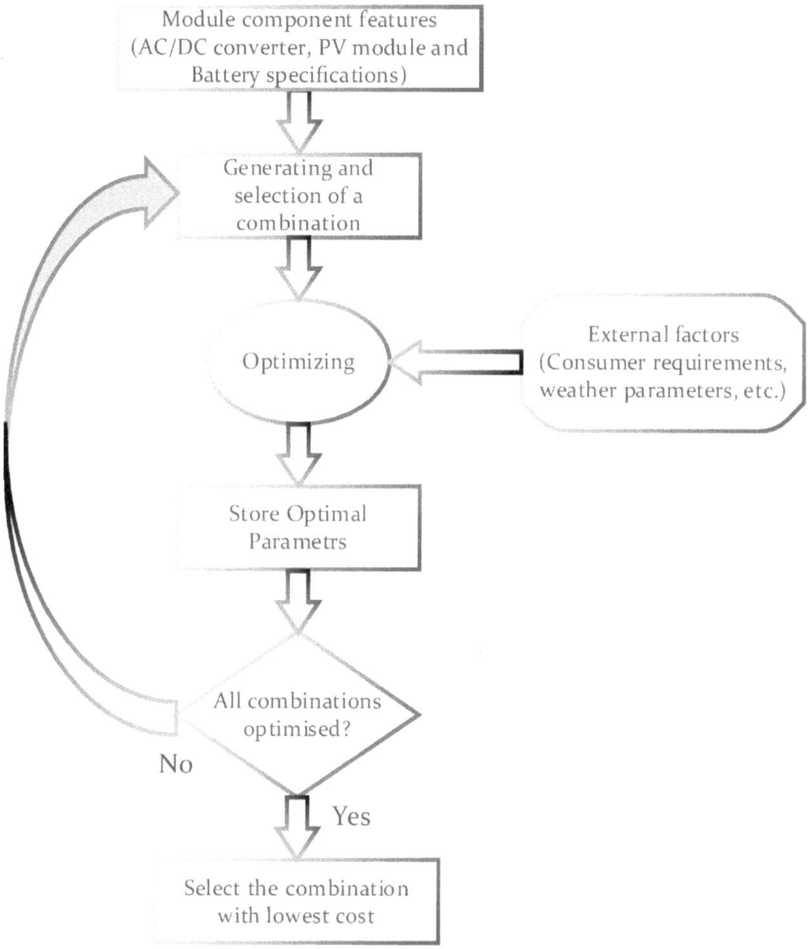

Figure 6.3 Flowchart for optimizing, sizing, and configuration of a PV plant.

4. Grid management

Users/stakeholders demand more reliable and successful grid maintenance and control as rapid developments in the power grid begin to make it smarter. When more players get involved in the power grid, maintaining such a large network gets more difficult. As a result, intelligent techniques are needed to cater to better smart grid management. In this segment, some of the issues that power grids face such as balancing supply and demand, grid operations and maintenance, management of grid data are introduced and the potential ML algorithm solutions to these issues are discussed.

4.1 Balancing both demand and supply

With the rising presence of renewable energy in the power system and the expanding diversity of loads, energy management has become complex due to the unpredictability of load demand and supply from renewable energy resources. Accurate load and supply predictions would assist in achieving a supply—demand balance, resulting in the more effective use of the residential grid.

Different ML models such as SVM, MLP, and LSTM are investigated in Ref. 74 to predict the PV generation and load demand using real-world data sets. Authors conclude that the MLP model produced the best results followed by the LSTM for load prediction, and for the prediction of PV power, the LSTM model produced the best results.

A ML model using SVM was developed in Ref. 75 to predict renewable energy generation 1 day ahead. Because of the prediction model's inaccuracy, it shows various uncertainties, which are characterized as uncontrolled base loads and excess renewable energy.

4.2 Operations and maintenance of the grid

Rudin et al.[76] gave a summary of ML approaches utilized in the power system for grid operations and management. The system was made up of several models for predicting various stages of the whole grid. These models may be used straight by electricity providers to help with maintenance and repair job prioritizing; (1) feeder failure rankings, and (2) cable, joint, termination, and transformer rankings. The authors employed roughly 300 features produced from time-series data and related parameters in their model. Models were built using CART, SVM, and ensemble learning approaches such as statistical methods and random forests. Using ML and Gaussian process regression,[77] created an Energy Management Model that combines smart grids and RESs. This model was created to address the issue of energy demand management.

4.3 Grid data management

There are numerous advantages to compressing power plant data sets. Compressed data sets may be easily shared by the academic and industrial communities; data storage costs, as well as the costs of monitoring the stored data, are reduced; and the data may be more accessible to data mining. This might result in the development of conveniently available extended resource data sets as well as the capacity to precisely and totally reproduce historical data for postmortem studies.

The author[78] demonstrated a lossless encoding of data of wind plants using its characteristics. They propose two approaches for the use of grid-connected wind farms and unordered wind farms. The authors believed that their methods were superior, with 50% more compression than other methods and a compression rate of 1.8 times. For example, a 1.5 compression rate indicates that 1.5 MB of data is compressed to 1 MB of data, requiring the user to deal with just 1 MB of data rather than 1.5 MB of data.

In Ref. 79, the authors proposed a compression technique for big-time series data sets. The authors developed a technique that employed filtered spikes, FFT, and projection of random signals to reflect the base data. This method gave a data reduction score of 91.5%. The technique used was lossy compression, but the essential information was still retained.

5. Maximum power point tracking

The energy production of a PV panel is affected by changes in environmental conditions. The conversion efficiency of a PV system is affected by variations in fog, heat, and cloud cover. Dust and other contaminants in the air or on the panel may also significantly reduce the performance of the power conversion operation. Thus, many research efforts are devoted to determining how to put PV panels to their full production capacity under any weather environment, thus safeguarding their profitability. Scientists determined that for a particular set of solar irradiation and temperature, there exists a precise operating point when the PV power output is at its peak. Any variation in output voltage caused by changes in load would result in the PV panel generating less than the maximum power. Temperature and solar irradiation have an impact on electricity generation as well. The goal is to identify an operating point where the maximum power point of the PV panel matches the load needs.

Several techniques are present to attain the maximum power in PV systems. Perturb & observe, and incremental conductance algorithms were two popular ways to solve the challenge. Because of the simplicity with which they can be implemented, both approaches are commonly used. They do, however, have disadvantages, such as instability due to vulnerability to sudden changes in weather conditions and oscillations around the operating point. Therefore, ML techniques are needed to overcome all the above disadvantages.

Two ML algorithms, namely, GRNN and SVR are utilized in Ref. 80 to reliably forecast the ideal reference voltage of the PV panel. A PI controller is used with the above voltage as a reference value such that the DC–DC converter will have constant output voltage and max. power even under variable environmental and load circumstances. In Ref. 81, authors suggested an algorithm for MPPT that is based on reinforcement learning.

6. Conclusion

The traditional energy sources are declining day by day and also burning fossil fuels causes environmental pollution. Therefore, governments are increasingly concentrating on utilizing RES. Hydropower is a perfect representation of a clean energy source that has been effectively employed for years. Wind and solar are two other auspicious RES, those have seen rapid growth in recent years. The energy production capabilities of these resources are not totally predictable or controllable, demanding the employment of appropriate management and forecasting approaches to enable seamless integration into the grid. In order to anticipate sudden variations in power supply and demand, the grid that integrates RES must be continuously observed. The research presented in this paper looks at how ML can be used to boost renewable energy generation and, more specifically, how it can be integrated into the power grid. ML algorithms for any characteristic of renewable energy generation and grid integration are difficult to generalize, but good collaboration across the many predictions and decision-making models is required to improve the overall performance and usefulness of the grid. Furthermore, ML approaches have been effectively employed to design renewable energy plants with reasonable precision based on existing data. The availability of published literature based on ML techniques for locating, sizing, and configuring wind and PV installations demonstrates their popularity, particularly in distant places. This highlights ML's potential as a modeling approach for renewable energy strategic planning and policy creation.

References

1. Ibrahim MS, Dong W, Yang Q. Machine learning driven smart electric power systems: current trends and new perspectives. *Appl Energy*. 2020:272. https://doi.org/10.1016/j.apenergy.2020.115237.
2. Lai JP, Chang YM, Chen CH, Pai PF. A survey of machine learning models in renewable energy predictions. *Appl Sci*. 2020;10(17). https://doi.org/10.3390/app10175975.
3. Olabi AG. Renewable energy and energy storage systems. *Energy*. 2017;136:1–6. https://doi.org/10.1016/j.energy.2017.07.054.
4. Zendehboudi A, Baseer MA, Saidur R. Application of support vector machine models for forecasting solar and wind energy resources: a review. *J Clean Prod*. 2018;199:272–285. https://doi.org/10.1016/j.jclepro.2018.07.164.

5. Tripathy DS, Prusty BR. *Forecasting of Renewable Generation for Applications in Smart Grid Power Systems*. Elsevier BV; 2021:265–298. https://doi.org/10.1016/b978-0-12-824337-4.00010-2.
6. Shafiullah PW, Amanullah MT, Oo D, Jarvis S. Potential challenges: integrating renewable energy with the smart grid. In: *Power Quality for the 21st Century Proceedings Australasian Universities Power Enginering 2010 Conference*. Vol. 6. 2010.
7. Belaid S, Mellit A, Boualit H, Zaiani M. Hourly global solar forecasting models based on a supervised machine learning algorithm and time series principle. *Int J Ambient Energy*. 2020. https://doi.org/10.1080/01430750.2020.1718754.
8. Bamisile O, Oluwasanmi A, Obiora S, Osei-Mensah E, Asoronye G, Huang Q. Application of deep learning for solar irradiance and solar photovoltaic multi-parameter forecast. *Energy Sources, Part A Recovery, Util Environ Eff.* 2020. https://doi.org/10.1080/15567036.2020.1801903.
9. Yildiz C, Acikgoz H. A kernel extreme learning machine-based neural network to forecast very short-term power output of an on-grid photovoltaic power plant. *Energy Sources Part A Recovery Util Environ Eff.* 2021;43(4):395–412. https://doi.org/10.1080/15567036.2020.1801899.
10. Bendali W, Saber I, Boussetta M, Mourad Y, Bourachdi B, Bossoufi B. Deep learning for very short term solar irradiation forecasting. In: *2020 5th International Conference on Renewable Energies for Developing Countries, REDEC 2020*. 2020. https://doi.org/10.1109/REDEC49234.2020.9163897.
11. Parra Raffán LC, Romero A, Martinez M. Solar energy production forecasting through artificial neuronal networks, considering the Föhn, north and south winds in San Juan, Argentina. *J Eng*. 2019;2019(18):4824–4829. https://doi.org/10.1049/joe.2018.9368.
12. Hassan MZ, Ali MEK, Ali ABMS, Kumar J. Forecasting day-ahead solar radiation using machine learning approach. In: *Proceedings - 2017 4th Asia-Pacific World Congress on Computer Science and Engineering, APWC on CSE 2017*. 2018:252–258. https://doi.org/10.1109/APWConCSE.2017.00050.
13. Zang H, Cheng L, Ding T, Cheung KW, Wei Z, Sun G. Day-ahead photovoltaic power forecasting approach based on deep convolutional neural networks and meta learning. *Int J Electr Power Energy Syst*. 2020;118. https://doi.org/10.1016/j.ijepes.2019.105790.
14. Ağbulut Ü, Gürel AE, Biçen Y. Prediction of daily global solar radiation using different machine learning algorithms: evaluation and comparison. *Renew Sustain Energy Rev*. 2021;135. https://doi.org/10.1016/j.rser.2020.110114.
15. Benhmed K, Touati F, Al-Hitmi M, et al. PV power prediction in Qatar based on machine learning approach. In: *Proceedings of 2018 6th International Renewable and Sustainable Energy Conference, IRSEC 2018*. 2018. https://doi.org/10.1109/IRSEC.2018.8702880.
16. Burianek T, Misak S. Solar irradiance forecasting model based on extreme learning machine. In: *EEEIC 2016 - International Conference on Environment and Electrical Engineering*. 2016. https://doi.org/10.1109/EEEIC.2016.7555445.
17. Ağbulut Ü, Gürel AE, Ergün A, Ceylan İ. Performance assessment of a V-trough photovoltaic system and prediction of power output with different machine learning algorithms. *J Clean Prod*. 2020:268. https://doi.org/10.1016/j.jclepro.2020.122269.
18. Üstün I, Üneş F, Mert I, Karakuş C. *A Comparative Study of Estimating Solar Radiation Using Machine Learning Approaches: DL, SMGRT, and ANFIS. Energy Sources, Part A: Recovery, Utilization and Environmental Effects*. 2020. https://doi.org/10.1080/15567036.2020.1781301.
19. Puah BK, Chong LW, Wong YW, et al. A regression unsupervised incremental learning algorithm for solar irradiance prediction. *Renew Energy*. 2021;164:908–925. https://doi.org/10.1016/j.renene.2020.09.080.

20. Afrasiabi M, Mohammadi M, Rastegar M, Afrasiabi S. Deep learning architecture for direct probability density prediction of small-scale solar generation. *IET Gener, Transm Distrib*. 2020;14(11):2017–2025. https://doi.org/10.1049/iet-gtd.2019.1289.
21. Wang J, Guo L, Zhang C, Song L, Duan J, Duan L. Thermal power forecasting of solar power tower system by combining mechanism modeling and deep learning method. *Energy*. 2020:208. https://doi.org/10.1016/j.energy.2020.118403.
22. Alizamir M, Kim S, Kisi O, Zounemat-Kermani M. A comparative study of several machine learning based non-linear regression methods in estimating solar radiation: case studies of the USA and Turkey regions. *Energy*. 2020;197. https://doi.org/10.1016/j.energy.2020.117239.
23. Koo C, Li W, Cha SH, Zhang S. A novel estimation approach for the solar radiation potential with its complex spatial pattern via machine-learning techniques. *Renew Energy*. 2019;133:575–592. https://doi.org/10.1016/j.renene.2018.10.066.
24. Javed A, Kasi BK, Khan FA. Predicting solar irradiance using machine learning techniques. In: *2019 15th International Wireless Communications and Mobile Computing Conference, IWCMC 2019*. 2019:1458–1462. https://doi.org/10.1109/IWCMC.2019.8766480.
25. Tripathy DS, Prusty BR, Jena D. Short-term PV generation forecasting using quantile regression averaging. In: *2020 IEEE International Conference on Power Systems Technology, Powercon 2020*. 2020. https://doi.org/10.1109/POWERCON48463.2020.9230535.
26. Tripathy DS, Prusty BR, Jena D. Probabilistic forecasting of daily PV generation using quantile regression method. In: *Proceedings—2020 IEEE India Council International Subsections Conference, INDISCON 2020*. 2020:260–265. https://doi.org/10.1109/INDISCON50162.2020.00060.
27. Feng C, Cui M, Hodge BM, Zhang J. A data-driven multi-model methodology with deep feature selection for short-term wind forecasting. *Appl Energy*. 2017;190:1245–1257. https://doi.org/10.1016/j.apenergy.2017.01.043.
28. Lahouar A, Ben Hadj Slama J. Hour-ahead wind power forecast based on random forests. *Renew Energy*. 2017;109:529–541. https://doi.org/10.1016/j.renene.2017.03.064.
29. Hong YY, Satriani TRA. Day-ahead spatiotemporal wind speed forecasting using robust design-based deep learning neural network. *Energy*. 2020;209. https://doi.org/10.1016/j.energy.2020.118441.
30. Chaudhary A, Sharma A, Kumar A, Dikshit K, Kumar N. Short term wind power forecasting using machine learning techniques. *J Stat Manag Syst*. 2020;23(1):145–156. https://doi.org/10.1080/09720510.2020.1721632.
31. Zhang J, Meng H, Gu B, Li P. Research on short-term wind power combined forecasting and its Gaussian could uncertainty to support the integration of renewables and EVs. *Renew Energy*. 2020;153:884–899. https://doi.org/10.1016/j.renene.2020.01.062.
32. Ali MEK, Hassan MZ, Ali ABMS, Kumar J. Prediction of wind speed using real data: an analysis of statistical machine learning techniques. In: *Proceedings—2017 4th Asia-Pacific World Congress on Computer Science and Engineering, APWC on CSE 2017*. 2018:259–264. https://doi.org/10.1109/APWConCSE.2017.00051.
33. Mahmoud T, Dong ZY, Ma J. Advanced method for short-term wind power prediction with multiple observation points using extreme learning machines. *J Eng*. 2018;2018(1):29–38. https://doi.org/10.1049/joe.2017.0338.
34. Zhang Y, Wang P, Zhang C, Lei S. Wind energy prediction with LS-SVM based on Lorenz perturbation. *J Eng*. 2017;2017(13):1724–1727. https://doi.org/10.1049/joe.2017.0626.

35. Li Y, Yang P, Wang H. Short-term wind speed forecasting based on improved ant colony algorithm for LSSVM. *Cluster Comput.* 2019;22:11575–11581. https://doi.org/10.1007/s10586-017-1422-2.
36. Manero J, Béjar J, Cortés U. "Dust in the wind...", deep learning application to wind energy time series forecasting. *Energies.* 2019;12(12). https://doi.org/10.3390/en12122385.
37. Mujeeb S, Javaid N, Gul H, Daood N, Shabbir S, Arif A. Wind power forecasting based on efficient deep convolution neural networks. In: *Lecture Notes in Networks and Systems.* Vol. 96. Pakistan: Springer; 2020:47–56. https://doi.org/10.1007/978-3-030-33509-0_5.
38. Jiao R, Huang X, Ma X, Han L, Tian W. A model combining stacked Auto encoder and back propagation algorithm for short-term wind power forecasting. *IEEE Access.* 2018;6:17851–17858. https://doi.org/10.1109/ACCESS.2018.2818108.
39. Huang CJ, Kuo PH. A short-term wind speed forecasting model by using artificial neural networks with stochastic optimization for renewable energy systems. *Energies.* 2018; 11(10). https://doi.org/10.3390/en11102777.
40. Eze EC, Chatwin CR, Brighton S. Enhanced recurrent neural network for short-term wind farm power output prediction. *J Appl Sci.* 2019:28–35.
41. Dong D, Sheng Z, Yang T. Wind power prediction based on recurrent neural network with long short-term memory units. In: *2018 IEEE International Conference on Renewable Energy and Power Engineering, REPE 2018.* 2019:34–38. https://doi.org/10.1109/REPE.2018.8657666.
42. Wang Z, Wang B, Liu C, Wang W. Improved BP neural network algorithm to wind power forecast. *J Eng.* 2017;2017(13):940–943. https://doi.org/10.1049/joe.2017.0469.
43. Aksoy B, Selbaş R. Estimation of wind turbine energy production value by using machine learning algorithms and development of implementation program. *Energy Sources, Part A Recovery, Util Environ Eff.* 2021;43(6):692–704. https://doi.org/10.1080/15567036.2019.1631410.
44. Sharifzadeh M, Sikinioti-Lock A, Shah N. Machine-learning methods for integrated renewable power generation: a comparative study of artificial neural networks, support vector regression, and Gaussian Process Regression. *Renew Sustain Energy Rev.* 2019; 108:513–538. https://doi.org/10.1016/j.rser.2019.03.040.
45. Harbola S, Coors V. One dimensional convolutional neural network architectures for wind prediction. *Energy Convers Manag.* 2019;195:70–75. https://doi.org/10.1016/j.enconman.2019.05.007.
46. Scene learning : deep convolutional networks for wind power prediction by embedding turbines into grid space. *Appl Energy.* 2019;238:249–257. https://doi.org/10.1016/j.apenergy.2019.01.010.
47. Demolli H, Dokuz AS, Ecemis A, Gokcek M. Wind power forecasting based on daily wind speed data using machine learning algorithms. *Energy Convers Manag.* 2019;198. https://doi.org/10.1016/j.enconman.2019.111823.
48. Zhang Z, Qin H, Liu Y, et al. Wind speed forecasting based on quantile regression minimal gated memory network and kernel density estimation. *Energy Convers Manag.* 2019;196:1395–1409. https://doi.org/10.1016/j.enconman.2019.06.024.
49. Sapitang M, Ridwan WM, Kushiar KF, Ahmed AN, El-Shafie A. Machine learning application in reservoir water level forecasting for sustainable hydropower generation strategy. *Sustainability.* 2020;12(15). https://doi.org/10.3390/su12156121.
50. Guo T, He W, Jiang Z, Chu X, Malekian R, Li Z. An improved LSSVM model for intelligent prediction of the daily water level. *Energies.* 2019;12(1). https://doi.org/10.3390/en12010112.

51. Üneş F, Demirci M, Taşar B, Kaya YZ, Varçin H. Estimating dam reservoir level fluctuations using data-driven techniques. *Pol J Environ Stud.* 2019;28(5):3451–3462. https://doi.org/10.15244/pjoes/93923.
52. Zhang S, Lu L, Yu J, Zhou H. Short-term water level prediction using different artificial intelligent models. In: *2016 5th International Conference on Agro-Geoinformatics, Agro-Geoinformatics 2016.* 2016. https://doi.org/10.1109/Agro-Geoinformatics.2016.7577678.
53. Kavousi-Fard A, Su W. A combined prognostic model based on machine learning for tidal current prediction. *IEEE Trans Geosci Rem Sens.* 2017;55(6):3108–3114. https://doi.org/10.1109/TGRS.2017.2659538.
54. Gangwani P, Soni J, Upadhyay H, Joshi S. A deep learning approach for modeling of geothermal energy prediction. *Int J Comput Sci Inf Secur.* 2020;18:62–65.
55. Assouline D, Mohajeri N, Gudmundsson A, Scartezzini JL. A machine learning approach for mapping the very shallow theoretical geothermal potential. *Geoth Energy.* 2019;7(1). https://doi.org/10.1186/s40517-019-0135-6.
56. Elmaz F, Yücel Ö, Mutlu AY. Predictive modeling of biomass gasification with machine learning-based regression methods. *Energy.* 2020:191. https://doi.org/10.1016/j.energy.2019.116541.
57. Samadi SH, Ghobadian B, Nosrati M. Prediction of higher heating value of biomass materials based on proximate analysis using gradient boosted regression trees method. *Energy Sources Part A Recovery Util Environ Eff.* 2021;43(6):672–681. https://doi.org/10.1080/15567036.2019.1630521.
58. Ozbas EE, Aksu D, Ongen A, Aydin MA, Ozcan HK. Hydrogen production via biomass gasification, and modeling by supervised machine learning algorithms. *Int J Hydrog Energy.* 2019;44(32):17260–17268. https://doi.org/10.1016/j.ijhydene.2019.02.108.
59. Mellit A. Sizing of photovoltaic systems: a review. *Rev des Energies Renouvelables.* 2007; 10:463–472.
60. Yokoyama R, Ito K, Yuasa Y. Multiobjective optimal unit sizing of hybrid power generation systems utilizing photovoltaic and wind energy. *Journal of Solar Energy Engineering, Transactions of the ASME.* 1994;116(4):167–173. https://doi.org/10.1115/1.2930078.
61. Mellit A, Benghanem M, Arab AH, Guessoum A. Modelling of sizing the photovoltaic system parameters using artificial neural network. *IEEE Conference on Control Applications - Proceedings.* 2003;1:353–357.
62. Mellit A, Benghanem M, Arab AH, Guessoum A, Moulai K. Neural network adaptive wavelets for sizing of stand-alone photovoltaic systems. In: *2004 2nd International IEEE Conference "Intelligent Systems" - Proceedings.* Vol. 1. 2004:365–370.
63. Senjyu T, Hayashi D, Yona A, Urasaki N, Funabashi T. Optimal configuration of power generating systems in isolated island with renewable energy. *Renew Energy.* 2007; 32(11):1917–1933. https://doi.org/10.1016/j.renene.2006.09.003.
64. Hernández JC, Medina A, Jurado F. Optimal allocation and sizing for profitability and voltage enhancement of PV systems on feeders. *Renew Energy.* 2007;32(10): 1768–1789. https://doi.org/10.1016/j.renene.2006.11.003.
65. Azadeh A, Sheikhalishahi M, Asadzadeh SM. A flexible neural network-fuzzy data envelopment analysis approach for location optimization of solar plants with uncertainty and complexity. *Renew Energy.* 2011;36(12):3394–3401. https://doi.org/10.1016/j.renene.2011.05.018.
66. Gurubel KJ, Osuna-Enciso V, Cardenas JJ, Coronado-Mendoza A, Perez-Cisneros MA, Sanchez EN. Neural forecasting and optimal sizing for hybrid renewable energy systems with grid-connected storage system. *J Renew Sustain Energy.* 2016;8(4). https://doi.org/10.1063/1.4960125.

67. Martínez M, Molina MG, Mercado PE. Optimal sizing method of vanadium redox flow battery to provide load frequency control in power systems with intermittent renewable generation. *IET Renew Power Gener.* 2017;11(14):1804–1811. https://doi.org/10.1049/iet-rpg.2016.0798.
68. Ali ES, Abd Elazim SM, Abdelaziz AY. Optimal allocation and sizing of renewable distributed generation using ant lion optimization algorithm. *Electr Eng.* 2018;100(1): 99–109. https://doi.org/10.1007/s00202-016-0477-z.
69. De La Torre S, González-González JM, Aguado JA, Martín S. Optimal battery sizing considering degradation for renewable energy integration. *IET Renew Power Gener.* 2019;13(4):572–577. https://doi.org/10.1049/iet-rpg.2018.5489.
70. Anand P, Rizwan M, Bath SK. Sizing of renewable energy based hybrid system for rural electrification using grey wolf optimisation approach. *IET Energy Systems Integration.* 2019;1(3):158–172. https://doi.org/10.1049/iet-esi.2018.0053.
71. Mahmoud K, Abdel-Nasser M. Fast yet accurate energy-loss-assessment approach for analyzing/sizing PV in distribution systems using machine learning. *IEEE Trans Sustain Energy.* 2019;10(3):1025–1033. https://doi.org/10.1109/TSTE.2018.2859036.
72. Quiles E, Roldán-Blay C, Escrivá-Escrivá G, Roldán-Porta C. Accurate sizing of residential stand-alone photovoltaic systems considering system reliability. *Sustainability.* 2020;12(3). https://doi.org/10.3390/su12031274.
73. Perera KS, Aung Z, Woon WL. Machine learning techniques for supporting renewable energy generation and integration: a survey. Lecture notes in computer science (including subseries lecture notes in artificial intelligence and lecture notes in bioinformatics). In: *International Workshop on Data Analytics for Renewable Energy Integration.* 8817. 2014:81–96. https://doi.org/10.1007/978-3-319-13290-7_7.
74. Golder A, Jneid J, Zhao J, Bouffard F. Machine learning-based demand and PV power forecasts. In: *2019 IEEE Electrical Power and Energy Conference, EPEC 2019.* 2019. https://doi.org/10.1109/EPEC47565.2019.9074819.
75. Minhas DM, Khalid RR, Frey G. Load control for supply-demand balancing under Renewable Energy forecasting. In: *2017 IEEE 2nd International Conference on Direct Current Microgrids, ICDCM 2017.* 2017:365–370. https://doi.org/10.1109/ICDCM.2017.8001071.
76. Rudin C, Waltz D, Anderson R, et al. Machine learning for the New York City power grid. *IEEE Trans Pattern Anal Mach Intell.* 2012;34(2):328–345. https://doi.org/10.1109/TPAMI.2011.108.
77. Ahmed W, Ansari H, Khan B, et al. Machine learning based energy management model for smart grid and renewable energy districts. *IEEE Access.* 2020;8:185059–185078. https://doi.org/10.1109/ACCESS.2020.3029943.
78. Louie H, Miguel A. Lossless compression of wind plant data. *IEEE Trans Sustain Energy.* 2012;3(3):598–606. https://doi.org/10.1109/TSTE.2012.2195039.
79. Reeves G, Liu J, Nath S, Zhao F. Managing massive time series streams with multi-scale compressed trickles. *Proc VLDB Endow.* 2009;2(1):97–108. https://doi.org/10.14778/1687627.1687639.
80. Takruri M, Farhat M, Barambones O, et al. Maximum power point tracking of PV system based on machine learning. *Energies.* 2020;13(3). https://doi.org/10.3390/en13030692.
81. Kofinas P, Doltsinis S, Dounis AI, Vouros GA. A reinforcement learning approach for MPPT control method of photovoltaic sources. *Renew Energy.* 2017;108:461–473. https://doi.org/10.1016/j.renene.2017.03.008.

CHAPTER SEVEN

Role of big data analytic and machine learning in power system contingency analysis

Ravi V. Angadi[1], Suresh Babu Daram[2], P.S. Venkataramu[1]
[1]Presidency University, Bengaluru, Karnataka, India
[2]Department of EEE, Centre for Energy, Sree Vidyanikethan Engineering College, Tirupati, Andra Pradesh, India

Contents

Smart Electrical and Mechanical Systems
ISBN: 978-0-323-90789-7
https://doi.org/10.1016/B978-0-323-90789-7.00004-X

1. Introduction

The power system is the most complex human-made system in the world. Power engineers face difficult challenges to manage and control the power flow in the system. The interconnected operation adds more complexity to the system. The main goal of a well-functioning power system is to supply electrical energy to consumers at the required voltage and frequency on a continuous basis. To achieve the goal, power systems should be operated at high levels of protection, with sufficient redundancy and reserve margins in generation and transmission systems. These security thresholds must be tracked in real time in order to take any corrective or protective steps necessary to ensure the system's smooth running. The operation of power system network is operating under stress condition largely because of the constantly rising demand of load, decreasing energy supplies and environmental limits on the extension of the transmission line. The electric power networks have undergone several difficulties and technical advances in recent years and are digitalized by the first smart grid ideas. The security monitoring in a modern energy management system (EMS) is a vital component of the system, but in real time it is a difficult work for the engineers of the power system.[1]

In an operations control center, system security is divided into two main functions, viz., security assessment and security control. The first reveals the current operating system's security level. The latter calculates the suitable security constrained scheduling required to achieve the desired security level optimally. Real-time and study modes are available for EMS security operations. The computational speed and dependability required by real-time application operations are very high. The static security level of a power system is defined by the existence or absence of emergency operating situations such as limit violation during pre- or postcontingency stages. System security is the procedure for detecting such ruptures. System evaluation generally consists of two roles, namely, system monitoring and contingency assessment. With the help of system monitoring, the power system operator may keep tabs on the present state of the power system. There are three separate stages to a contingency analysis: defining the contingency, selecting the contingencies, and evaluating the results.[2]

The contingency definition provides a list of possible scenarios that must be dealt with as soon as possible. A vast number of outages, such as branching or injecting outages, are included in this list. These risks are sorted from the most serious to the least serious using contingency selection methods to reduce the list. When speed is more important than accuracy, an approximation of the system model is employed. This is followed by a contingency review of the individual cases in decreasing severity, utilizing load flow. The evaluation procedure is carried on until no postcontingency ruptures are found. As a result, the goal of a contingency analysis is to determine the list of eventualities that, if true, would result in system operational state ruptures.[2] Due to the system contingency, the voltage variance will occur. The unit will pass through the critical contingency from protection to emergency operation mode. The severity of the lines in critical contingency thus needs to be predicted. During the static study analysis, a lot of simulation work to be carried out for the different operating conditions such as variations of load (under load, base load, and over load) and as well as for the various line outages or generator outage or transformer outage and so on. While performing the contingency analysis study, it will lead to high volume of data, these data need to handle effectively. The simulation data processing can be done easily with help of big data techniques.[3] The data handling will help to identify the valid data that will bring value to the system, and this processed data will help the system to identify the severity of the line.

The chapter comprises the following sections: Section 2 discusses traditional power system contingency analysis. Section 3 discusses the theory of computing voltage stability index and Section 4 discusses the applications of big data analytics in power system, application of machine learning was briefly described in Section 5. Section 6 discusses the proposed methodology. Case study and results are presented in Section 7. The last part of the chapter contains the conclusion.

2. Contingency analysis

An integrated electricity grid has already experienced several large-scale outages. In order to maintain reliable power delivery, the power system must be run at its most affordable. The ability to operate in a dependable manner requires that there is sufficient power generation and that this power can be successfully delivered to the loads. The majority of power systems are constructed with sufficient redundancy to ensure that they can withstand any significant failure event. Two significant failures, such as line outages and generating unit failure, have the potential to have severe effects and necessitate the implementation of corrective measures. In this chapter, a single line outage is taken into consideration in order to do contingency planning.

In order to analyze, to supply, and to prioritize the effects of electric power systems when issues emerge, contingency analyses are a "what if" scenario. A condition of this is the loss or failure of a minute portion of the energy system, such as a transmission line or equipment like a generator or a transformer. This is also known as an unintended failure or failure. In order to assess the consequences and to determine all overloads arising as consequence of all outage occurrences, the input testing is a computer program using a simulated power system model. The impacts of issues in the power system are modeled and measured in the short term. This chapter focused to analyze contingency occurrences off-line and to demonstrate operators the implications of future failures on planning recovery scenarios as an on-line tool.[4] A safety assessment of the power system is essential for the development of strategies in which one or more parts can maintain the system function. If one or more parts can endure the loss and continue operating without serious difficulties, a power system is secure. Contingency analysis is an application in the Power Utility Control Center, for security analysis, which distinguishes energy control from a less complicated SCADA system. It is to study the power system so that overloads and faults can be identified as a result of a contingency.[5]

If the power system is rebalanced after a contingency and does not overload a single part it is referred to be a normal condition, if numerous parts such as lines and transformers get overloaded, and if the power system becomes unstable and collapses as a critically important condition it risks harm. Fig. 7.1 shows the process of the events occurred during contingency condition. The existing operating regulations of electricity utilities (such as NERC) mandate that each power system should be able for every initial contingent or failure to endure and recover. Future policies might expand this to cover a second scenario or any other failure thereafter. Critical configurations can be detected, operational controls and restrictions can be implemented, and remedial measures can be prepared by preanalyzing the impact of contingency occurrences. The impacts of the device failures are assessed by future power system operating circumstances. This outage timetable requires modifications in the time, long-term and short-term planning

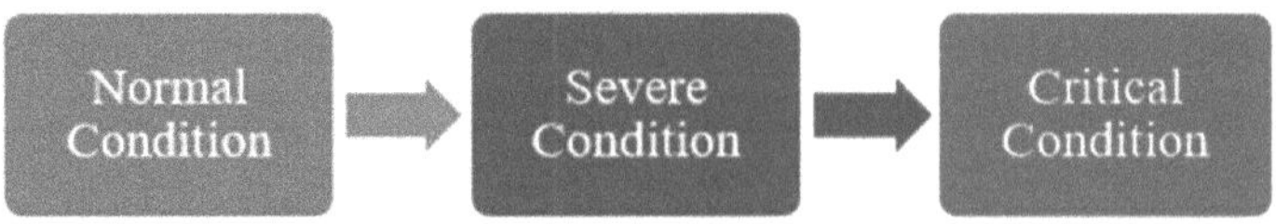

Figure 7.1 Flow of contingency analysis.

to reflect changes in working circumstances. Power system engineers, however, create the model with sufficient information to give the real power system an excellent simulation with precise results. Contingency analysis may be performed using a number of contingency occurrences with an initialized power network model.

Every element to be removed from a network model is included in a contingency list, to evaluate the impact of hypothetical overloads of other elements. The establishment of the contingency list is a consequence of planning trials using energy streams under different loading situations to identify the susceptible parts of the grid. Future enhancements to the human interface and establishment procedures might enhance the utilization of online contingency analysis. In a larger context, contingency analysis is used to analyze large power pools and broad networks for operational areas, such as Midwest Independent System Operator (ISO). Power marketing, trading companies ISO as well as Transmission System Operator (TSO) have driven several of these criteria. Further improvements are needed in terms of runtime, user tools and presentation of findings for these very large network models, to make the program efficient. In this chapter, condition number and line voltage stability index are used for contingency ranking. The condition number shall be calculated to investigate the severity of a line failure from the point of view of system stability. When calculated values are particularly sensitive to modest changes to input values, the phrasing of a problem is defined as incorrect.

The stability index of line voltage involves an active transmission line power flow. For each transmission line, the highest value of the line voltage stabilization index and the accompanying load increase in the bus results in system severity. Analysis of contingency is one of the most significant issues faced by mass power planning and operating engineers. A stable electricity supply is a key aspect of today's complicated societies. A power system is considered to be secured if it is safe or risk-free. Security is the system's capacity to resist a list of scenarios determined without any consequences.

2.1 An evaluation of the conventional power system's N-1 contingency

As a subset of power system security analysis, $N-1$ contingency analysis may be seen as a broader view, which is more generic in nature. Power system security evaluation, in addition to $N-1$ contingency analysis, incorporates additional critical components of security such as rotor angle stability, frequency stability, and voltage stability. Thus, security analysis establishes the power system's integrity and verifies the system's postcontingency status in terms of branch overloads and node voltage variations. The typical power system contingency analysis comprises many power-flow runs with a set of contingencies. The study is also known as N−1 security analysis since each power-flow run disables one component. A comparison is made between actual and permitted operating limits based on the bus voltage and branch currents that were obtained. As soon as one or a few restrictions are exceeded, the system is labeled as unreliable or insecure. When the EMS sends an alarm, TSOs take action based on the severity of the security threat. The primary goal of the corrective activity is to raise the degree of security in the electric power system to an acceptable level.[3] A power security analysis module is included in the modern EMS, which evaluates the degree of power system security on a regular basis and notifies the operator if the secure functioning of the power system is threatened. The implementation of the Newton−Raphson iterative technique for solving the set of nonlinear power-flow equations is the most common way for carrying out the above-described methodology. It is necessary to verify the operational limitations of network branches and nodes after resolving each contingency's power-flow equations as described in Ref. 3.

2.2 Violation types

Generally speaking, the most frequent types of contingencies are line contingency and generator contingency. There are two sorts of violations that are mostly caused by these circumstances, viz., low voltage violations and line MVA limits violations. Buses have low voltage violations; this indicates that the bus voltage is below the required value. Any bus voltage range is typically 0.95−1.05 p.u. Low voltage occurs when the voltage drops below 0.95 p.u. A high voltage issue occurs when the voltage climbs above 1.05 p.u. Reactive power is recognized to be the cause of voltage issues in the power system network. Hence, when a bus has low voltage, reactive power is given to raise the bus voltage profile. Buses absorb high voltage reactive power to maintain system normal voltage.

The system encounters line MVA limitations, when the line's MVA rating exceeds the specified value. This is due to the amplitude of the current increasing in that line. These lines are intended to withstand 125% of their MVA maximum. According to utility standards, a current of 80%−90% of the limit is considered alarming. This study will discuss various solutions to this problem.[6]

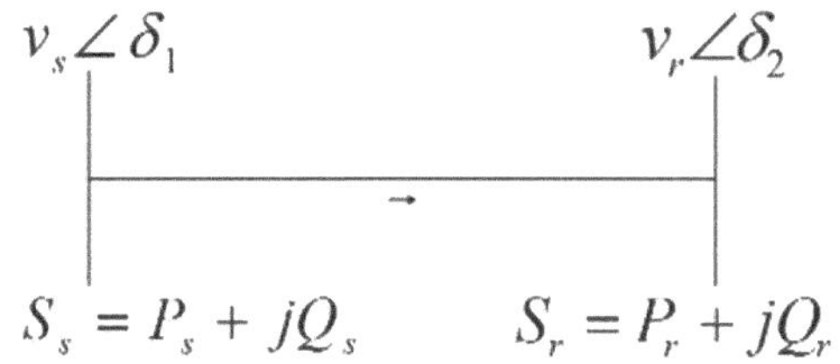

Figure 7.2 Single-line representation of two bus system.

2.3 Approach of contingency ranking

The convention of alternating current flow solution gives active, reactive power flows, and bus voltage magnitudes for contingency analysis. The line significance and the contingency ranking method of the power system were found. The ranking is produced by use of a voltage stabilization index of outcomes according to the severity calculation. The NR technique is used to analyze voltage stability indices utilizing each situation of contingency load and to examine NR.[7,8] Fig. 7.2 shows single-line diagram of two bus system.

where

$$L_{mn} = \frac{4XQ_r}{V_s Sin(\theta - \delta)^2} \leq 1.0 \tag{7.1}$$

Q_r = Receiving end reactive power in volt ampere reactive.
χ = Reactance of transmission line in ohms.
V_s = Voltage at sending end in volts.
θ = Impedance angle of line.
δ = Phase angle difference of Sending and Receiving end buses.

The contingency is determined and listed based on the resulting index, commencing with the maximum value of the Line Voltage Stability Index. The stability index of every line calculated in this work is a line voltage index for contingency analysis.[9] The constrained range is from zero to one for the system to be stable. If the value is bigger, the stronger the severity of the line.[7,8] If it is more likely, the more severe it is and the system is in the verge collapse.

3. Voltage stability analysis

Traditionally, the ability to transmit power has been restricted by either rotor angle stability or the capacity to withstand high heat loads.

A relationship has been established between the blackout problem and transient stability. Voltage stability, on the other hand, has become a primary concern in the design and operation of electric power networks and is a factor in limiting power transfers. Voltage stability refers to the capacity of the power system to maintain appropriate voltages across all busses in the system under normal and abnormal situations. When a disturbance causes a gradual and uncontrolled decrease in voltage, a power system is said to have reached a condition of voltage instability. Voltage instability or voltage collapse is caused by insufficient reactive power supply from generators and transmission lines, which has resulted in numerous significant system failures throughout the world.

As power systems become more stressed, the ability to maintain voltage stability becomes an increasing problem. Numerous occurrences of voltage collapse in utilities have brought the issue of voltage stability to the attention of power experts. Voltage stability impacts the whole power system, although it generally affects one crucial region. Voltage stability is the capacity of a power system to maintain stable acceptable voltages at all buses in the system under normal and abnormal situations. Many static indicators are utilized in the literature to assess the system's closeness to a static voltage collapse point. The condition number of a load flow Jacobian is one such index that produces a pretty accurate result and is utilized as a supporting technique in this thesis. This indicates the severity of a line outage in terms of system stability. An issue is ill-conditioned if the calculated values are highly sensitive to slight changes in the input values. The condition number measures the matrices sensitivity to a numerical operation.[4,7,8] In this work, LVSI was computed and the severity of the line was identified as described in Section 2.3. In the process of contingency analysis, a huge amount of data sets is involved. The handling of data effectively and the vital role of big data in power system is discussed in the below Section 4.

4. Applications of big data to power system

The issue of big data is critical in the context of power systems and also corresponds to the power system's application sector. Data are generated by a wide variety of digital sources, and the rapid adoption of digital technology has resulted in the rise of big data. The processing of massive data volumes is novel in a number of ways. This often entails collecting large or large-format data volumes that are challenging to manage using standard application or data processing tools. Structured, semistructured, or unstructured data sets can be acquired.[10] Big Data is technically classified as four Vs. Volume, Velocity, Variety, and Veracity are the four Vs depicted in Fig. 7.3.

The term "volume" refers to the massive amount of data created each day, and "speed" refers to the rate of growth and the rate at which data are gathered for analysis. Variety provides information about many data types, including structured, unstructured, and semistructured data. The fourth V stands for veracity, which encompasses both access and transparency. The primary goal of Big Data Analysis is to combine traditional and intelligent computational techniques in order to manage large volumes of long-term, flexible, and truthful data. The true importance of big data is not stated, but the belief is that it exists. It will enable us to make more informed decisions, demonstrate greater creativity, and optimize our operations while being creative and economically efficient.[10–12] Big Data is a topic which tackles ways to analyze, systemically extract, or otherwise treat massive or complicated data volumes that are handled by standard data processing programs. More row statistics have better statistical capacity; however, more complicated statistics lead to more erroneous findings. Data collection, storage, processing, sorting, distribution, transmission, streaming, assembling, updating, data conservation, and source data are all key issues.[10,13,25] Data must be analyzed using enhanced resources in order to investigate relevant information (analytics and algorithms).

4.1 Big data analytics process in power system

Big Data analytics examine large collections of usual unstructured knowledge in order to allow organizations to quicker find patterns, dynamics of the market and customer needs. Big Data Analytics for complicated insights,

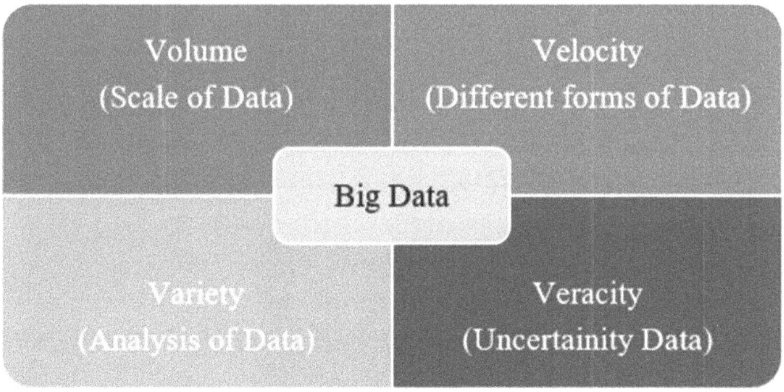

Figure 7.3 Characteristics of big data.

such as artificial intelligence and machine learning, can be employed for new technologies. The Big Data Analytics standard methods are data collection, processing, cleaning and analysis.[10] The method of Big Data processing is shown in Fig. 7.4.

4.2 Data collection

Before evaluating data, data must first be collected. This method may differ for the purpose of utilizing the data from one company to another. In this initial stage, three sort of data are collected.

Structured data: The acquired information is linear and structured in a relational database (think the data you see on a table). This kind of data is significantly simpler for large data analysis.

Unstructured data: approximately 80% of contemporary data is unstructuring, mostly created by human beings. This content is available via text, social media, videos, audio, and more. Big Data programs are harder to understand as unstructured data are various (and sometimes random).

Semistructured data: Semistructured data are a data type that cannot easily understand the machine language, but it has certain tagging information. This is examined in XML or e-Mail files. These forms of data can be chosen from cloud, server, and operational, embedded sensor, mobile, and other sources.

4.3 Data processing

It is ready to be processed and sorted for use after data has been collected and stored. But dealing with Big Data is a common challenge. Research from IDC shows that every 2 years the amount of data in the world is indeed

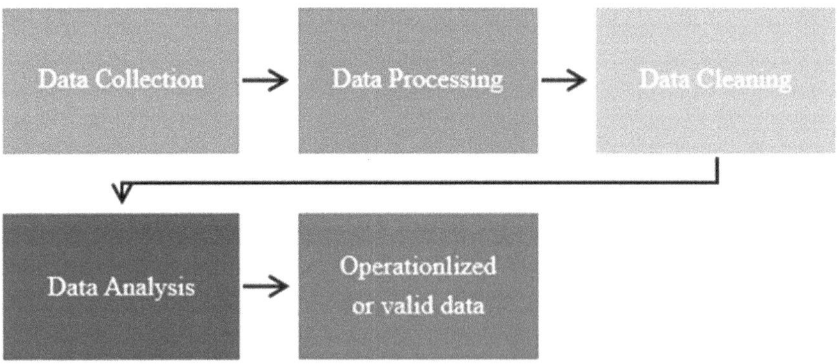

Figure 7.4 Big Data Analytics process flow in power system.

doubling—outweighing our ability to decipher it. This has led to increased processing in real time, but it is not the only way to process big data.

Batch processing: The batch processing is the time-relevant approach for the long-term collection of enormous amounts of data. Millions of bats can analyze data blocks, but their performance might take hours if not days.

Stream processing: Streaming is a more flexible approach for Big Data processing. If your company's success is driven by real-time analytics, stream processing is the way ahead. Stream processing can be beneficial for tasks like identifying fraudulent transactional data operations. This is a very sophisticated procedure and often costs more than batch processing.

Data preprocessing for data mining refers to the collection of procedures utilized prior to the deployment of a data mining method. Given the inherent imperfection of data, containing inconsistencies and redundancies is not a prerequisite for initiating a data mining process. Additionally, we must highlight the rapid increase in data creation rates and size in corporate, industrial, academic, and scientific applications.[14–17] Larger data sets necessitate more advanced data analysis methods. As a result of preprocessing, data that would otherwise be *unprocessable* can be processed. In the contingency analysis, the data generated from the load flow studies are huge and having different attributes, all these data need not be required to estimate the severity of the transmission line; hence, it is necessary process the data and identify the valid data that must require for the further process, this can be achieved with the data processing technique.

4.4 Data cleaning

The data obtained from multiple sources is messy, affecting the forecast result's accuracy. Data cleaning improves data quality, allowing the industry to prepare their data for analysis. For large data, most conventional approaches are no longer suitable. The data cleaning procedure finds and corrects mistakes. Despite the requirement to evaluate data rapidly, the data cleaning process is difficult and time intensive. Domain expertise is critical throughout the data cleansing process, as verification and validation are the primary issues for cleaned data. Any data to be utilized for analysis must be fundamentally standardized so that it is structured in the same way.[18] Data must also be appropriately migrated from outmoded or outdated systems or technologies no longer employed by a study. With the advancement of Big Data analysis technologies globally, the relevance of asset management for data analysis based on electric power facilities is rising. It is critical to safeguard the data quality that will decide the success of the asset management RISK assessment algorithm. Preprocessing asset data improves asset management dependability. It is important to create an algorithm to minimize processing time and enhance accuracy of filthy data.

4.5 Data analysis

The data acquired, saved, processed, and cleansed for consistence are now ready for evaluation. This last stage is how data scientists extract meaningful knowledge from big volumes and combinations—but not all analytical data are subject to the same picture.

4.6 Operationalized or valid data

Upon the data analysis using suitable analytical tools, a valid or operational data will be recognized for the next section, and the data will be provided for the machine learning tool to forecast the action of the plan in the upcoming operation for secure and stable system operation.[19]

In this chapter, the structure data were obtained from MATLAB simulation by creating contingency for different loading condition and these data were processed by using data analytic technique as required, to apply those data set to the machine learning tool to predict the severity of the line by following the steps (data collection, data processing, data cleaning, data analysis, and operationalized or valid data). This process will help the system to identify the important data set and brings value to the system operation. The data observed and obtained in this section will be fed to the machine learning as discussed in the next Section 5.

5. Machine learning in power system application

Machine learning methods used to address enormous amounts of distributed data in power system operations fall into two categories: open source and free software. The data analytics model may be taught to detect connections between data features and categories or values based on the information provided about the analyzed learning algorithms.[20] It is typically intended for the data analytics model to categorize probable classes among all

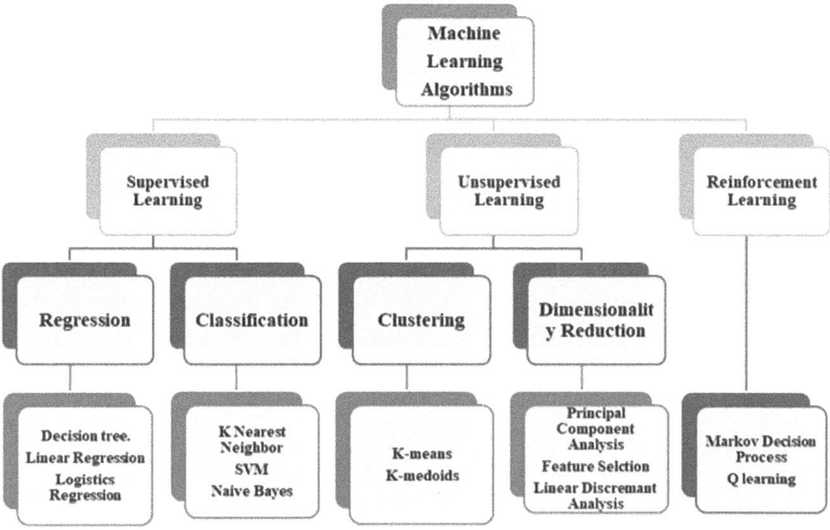

Figure 7.5 Classification of machine learning for Big Data.

items that do not have labels. According to Fig. 7.5, the most frequently used data extraction and machine learning algorithms are either supervised, unsupervised, and reinforcement learning methods, depending on whether each dimension is included in the data set.

In this chapter, the supervised learning (classification technique) is selected to carry out the proposed work, that is discussed in the next section.

5.1 Supervised learning

When using supervised learning, the computer is trained by a human. Using historical data, they can model future occurrences and make predictions based on what has already occurred. This circumstance necessitates the use of input and desired information to anticipate prospective outcomes. An error has been detected or the proper method has been identified in the data input. To begin, it should be remembered that all supervised learning algorithms are essentially complicated algorithms that may be divided into two categories: classification and regression models.

The techniques to machine learning that make predictions are frequently referred to as supervised learning. These methods of learning differ exactly

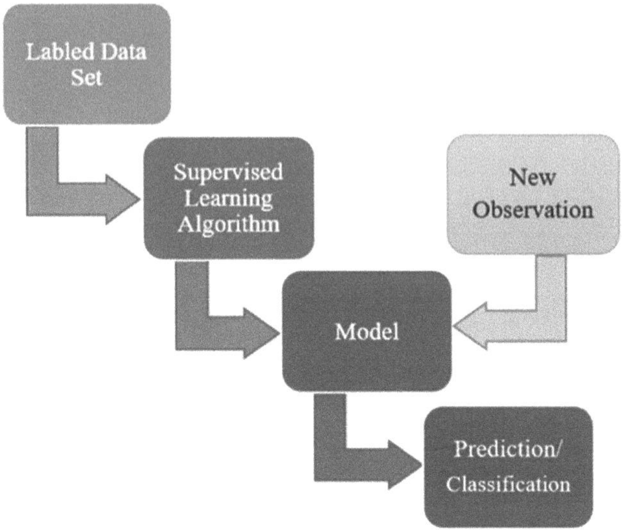

Figure 7.6 Process flow diagram of supervised learning.

from the unmonitored learning that models the distributions of cases in a classic one. The supervised learning approach is the approaches used which attempt to detect a link between the input and the target qualities.[16,20,21] In a structure called a model, the relationship which is identified is represented. The controlled approaches are grouped mostly into models like classification models and regression models. The process flow of overseeing learning is depicted in the preceding Fig. 7.6. In this chapter, the proposed work is analyzed by selecting the supervised learning, especially by selecting suitable classification algorithms technique to predict the severity of the transmission line.

5.2 Classification algorithm

In data classification, we divide our data into a desirable and distinct number of classes, with each class having a unique label that we may give to the data in it. Specifically, it is a data analysis task, which refers to the process of developing a model that defines and distinguishes between different data classes and ideas. Classes are subpopulations of observations that are classified into categories. Classification is the issue of determining which category an observation belongs to on the basis of a training set of data comprising observations and whose categories membership has been determined. Classification is a data mining method that uses training data to predict class labels and classify data.[21] This chapter discusses several categorization approaches that are frequently employed in the field of data mining and are used to determine the severity of a transmission line. The simulation generates a huge dataset with a variety of characteristics. These data will forecast the target variable for the new data set record when the j48 decision tree is used.[22,23] For the classification job, a decision tree is employed and it is sometimes called a tree of classification. The tree decision provides numerous supports for data mining, the end user can comprehend this quickly, number of inputs and text can be handled.[24–28] The decision classification tree, consisting of root node, all leaf nodes, and inner nodes describing requirements for examination is discussed.[16,20]

6. Proposed algorithm

In data mining, data are kept electronically and the search is automated—or at the very least enhanced— by a computer. This, too, is not novel. Economists, statisticians, forecasters, and communication engineers have long believed that patterns in data may be discovered, recognized, confirmed, and utilized for prediction. As the system becomes more complicated, overwhelming us in data, data mining becomes our only chance for illuminating hidden patterns. Data that have been intelligently evaluated is a useful resource. It has the potential to provide fresh insights and, in business contexts, competitive benefits.[16,21,22] Data mining is the process of addressing issues by evaluating data that has previously been collected in databases. The data mining approach was applied to the contingency analysis of the power system to predict the severity of the lines. The Lmn was calculated in accordance with Eq. (7.1). The following algorithm demonstrates the use of the constructed model; the model is shown to be effective.

Step No. 1: Read System Data.

Step No. 2: Compute the load flow solution by NR Method.

Step No. 3: Calculate Line voltage Stability (Lmn) values for different load conditions.

Step No. 4: Load the results into data base in.CSV file format.

Step. No. 5: Data processing using big data analytics tool.

Step. No. 6: Prepare the data with different attributes as per the contingency study.
Step. No. 7: Select Explorer in WEKA & Load the Data file data base.
Step. No. 8: Verify all the attributes are loaded or not.
Step No. 9: Click on the save option and save file in. ARFF format.
Step. No. 10: Click on classify and select classifier, choose Tree and select suitable classifier (say j48), and use training set and click on the start option & verify the run information in classifier output window.
Step. No. 11: Right click for options in the Result List window to view the visualize classifier errors, visualize tree, visualization margin curve, threshold curve, and visualize cost curve.
Step. No. 12: Click on visualize classifier errors and save the file to see the severity prediction of transmission line based on training data set.
Step. No. 13: Repeat the step 1 to 11, to predict the severity of the transmission line by suitable classification algorithm.
Step. No. 14: Print the severity of the transmission line for identified contingency and different load condition.

7. Case study and results

The IEEE-30 bus system is simulated to determine the suggested method's efficiency. The system's nominal MVA rating is 100 MVA. Convergence of active and reactive power is limited to 0.00,001. The IEEE-30 bus system consists of one slack bus, five generator buses, 24 load buses, and 41 transmission lines. For the purposes of this study, the single transmission line outage contingency condition is taken into account. Total of 32 contingencies are found in this scenario, and for each contingency, a certain transmission line is brought down, as shown in Table 7.1.

Table 7.2 indicates that top 10 contingency ranks for different load condition. From Table 7.2, it can be observed that the Lmn condition number increased with variation in system load. In this study, the contingency analysis was done by considering the single line transmission outage, each transmission line outage named with different contingency numbers. Example, if line two is outage under certain load condition, it is named as contingency number 1 (C. No. 1) and if line 19 was outage means it is named as contingency number 13, all the contingency are mentioned in Table 7.1. It is as clear that as the load varies, that is, either under load, base load, or over load under n-1 condition, the Lmn condition number also varied, it is mentioned in Table 7.2. It is observed that from Table 7.2, under load,

Table 7.1 List of contingency and number of transmission line considered for contingency.

Sl. No.	Contingency No.	Line number	Transmission line	
			From bus	To bus
1.	1.	2	1	3
2.	2.	3	2	4
3.	3.	4	3	4
4.	4.	5	2	5
5.	5.	6	2	6
6.	6.	7	4	6
7.	7.	8	5	7
8.	8.	9	6	7
9.	9.	10	6	8
10.	10.	14	9	10
11.	11.	17	12	14
12.	12.	18	12	15
13.	13.	19	12	16
14.	14.	20	14	15
15.	15.	21	16	17
16.	16.	22	15	18
17.	17.	23	18	19
18.	18.	24	19	20
19.	19.	25	10	20
20.	20.	26	10	17
21.	21.	27	10	21
22.	22.	28	10	22
23.	23.	29	21	22
24.	24.	30	15	23
25.	25.	31	22	24
26.	26.	32	23	24
27.	27.	33	24	25
28.	28.	35	25	27
29.	29.	37	27	29
30.	30.	38	27	29
31.	31.	39	29	30
32.	32.	40	8	28
	33.	41	6	28

base load, and over load condition, the transmission line number 12 claims the contingency rank number as 1, it means this particular line needs special attention. In the similar way, it is observed that during base load condition, transmission line number 22 is having contingency rank of 5 (C. No.1/Line No. 22/Lma 0.0571), during base load condition transmission line number

Table 7.2 Top 10 contingency ranking for different load condition.

	Under load condition	Base load condition	Over load condition
Rank No.	CN/Line/Lmn	CN/Line/Lmn	CN/Line/Lmn
1.	1/12/0.2476	1/12/0.3223	1/12/0.4379
2.	1/20/0.1119	1/22/0.2132	1/22/0.3606
3.	1/1/0.0763	1/20/0.1683	1/5/0.2966
4.	1/16/0.0651	1/5/0.1480	1/11/0.2683
5.	1/22/0.0571	1/16/0.1227	1/20/0.2373
6.	1/5/0.0452	1/1/0.1015	1/16/0.2145
7.	1/8/0.0426	1/11/0.0902	1/17/0.1990
8.	1/11/0.0353	1/3/0.0853	1/3/0.1807
9.	1/38/0.0316	1/17/0.0833	1/1/0.1118
10.	1/19/0.0300	1/33/0.0502	1/33/0.0823

16 is having contingency rank of 5 (C. No.1/Line No. 16/Lmn 0.1227), and during over load condition transmission line number 16 is having contingency rank of 5 (C. No.1/Line No. 20/Lmn 0.2373). Note: C. No one means transmission line two was outage.

In this process, it is required to continuous monitor of all the contingency along with variation in load conditions, which is required a special attention and which involves huge amount of data set, it so difficult to identify the crucial data. In order to support the contingency analysis study, the data mining technique will more suit and as well predict the severity of the transmission line as per the trained data sets. The similar analysis was carried for different contingencies along with different loading condition and listed top 10 contingency ranking is listed in the Table.2. The only contingency condition of the single transmission line is believed to follow the categorization approach in many circumstances. The ranking for the lines was based on this particular line on the line stability index value. The rankings are classified as below:

a. Critical condition: Ranks 1 to 10
b. Semicritical condition: Ranks 11 to 20
c. Noncritical condition: Ranks 21 to 40

The cases under consideration are as follows:

Case 1: Under load condition (Reduction of base load by 30%).

Case 2: Base load condition (100% load).

Case 3: Over load condition (Increase of base load by 30%).

The data are prepared separately by taking single transmission line outages into account, that is, distinct loading conditions, in all three scenarios.

Critical, semicritical, and noncritical ranges are categorized with classification algorithms. The produced data are firstly preprocessed and categorized as such. In this case first, it is identified that the various attributes such as Sl. No, Contingency Number, Line Number, Load condition, Line voltage stability index value (Lmn), and condition of the line from the simulation results, since the simulation results involve many attributes, which all are not essentially required to predict the severity of the line, so it is very much essential and need to process the data as per the required. Fig. 7.7 shows the various attributes that are uploaded to the Weka software machine learning tool. In the study, total 369 instance are uploaded with six attributes and sum of weights is 369. In this case, there are totally 34% of instance are missing that is observed in the simulation, and three distinct attributes in Contingency Number, Load condition, as well as in condition number, respectively. In line number attribute, there are total 41 distinct instances and 285 distinct value in the Lmn attributes.

The prediction of transmission line was achieved based on decision tree, which as 10 number of leaves and size of the tree is 17 as described in Fig. 7.8. The visualization of classifier window is shown in Fig. 7.9. It is evident that from the visualizer classifier it observed that it is predicted the severity condition for the missed instances. In order to analyze proposed method, different case studies are considered to predict severity of transmission line under different load condition and the contingencies are discussed below;

7.1 (Case 1): Under load condition (reduction of base load by 30%)

In this case, the LVSI index was computed for under load condition, the severity was predicted based on LVSI, the data generated during this process were resulted in high volume of data, and these data were processed by using data analytic and machine learning tools. In this case, 67% of the data is used for training to predict the severity of the line and 33% of the data is tested based on the training data set. The sample training data set for the underload condition is tabulated in Table 7.3 Based on these training data sets, severity of the other transmission line was predicted by using machine learning tool Weka, and tabulated in Table 7.4 along with various attributes.

7.2 (Case 2): Base load condition (100% load)

In this case, the analysis was carried out on actual load condition of the system. The LVSI index was computed for base load condition, the severity was predicted based on LVSI. In this case also, 67% simulation data is

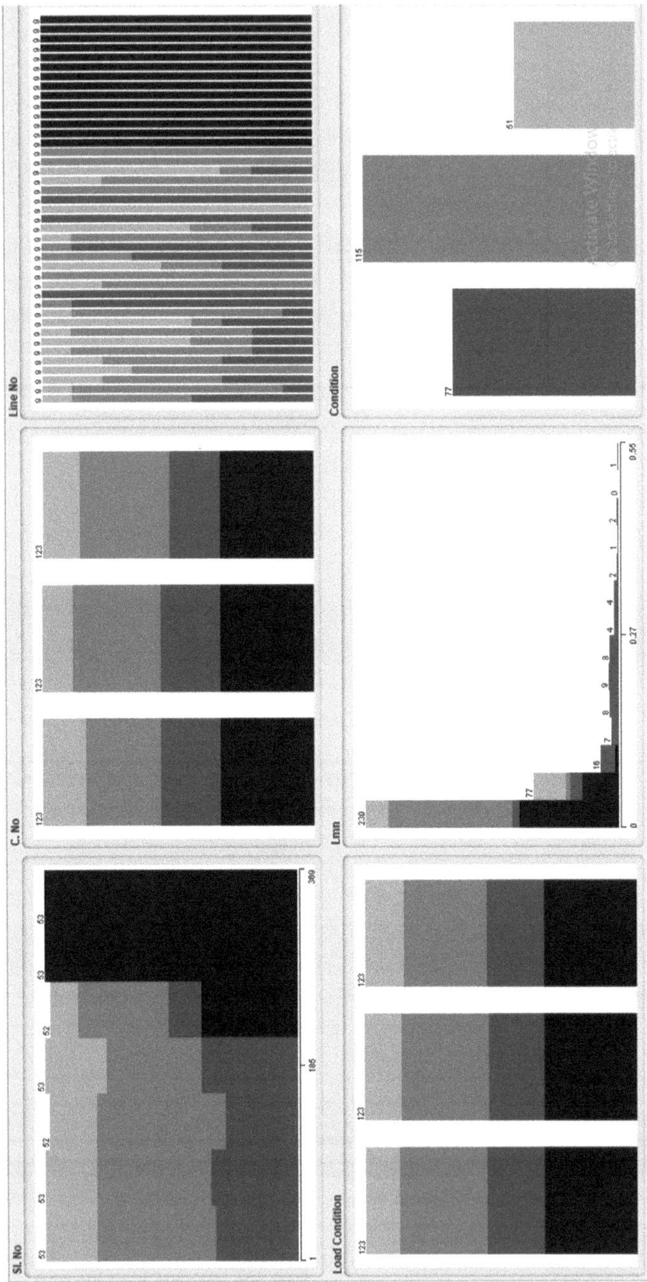

Figure 7.7 Data processing with various identified attributes.

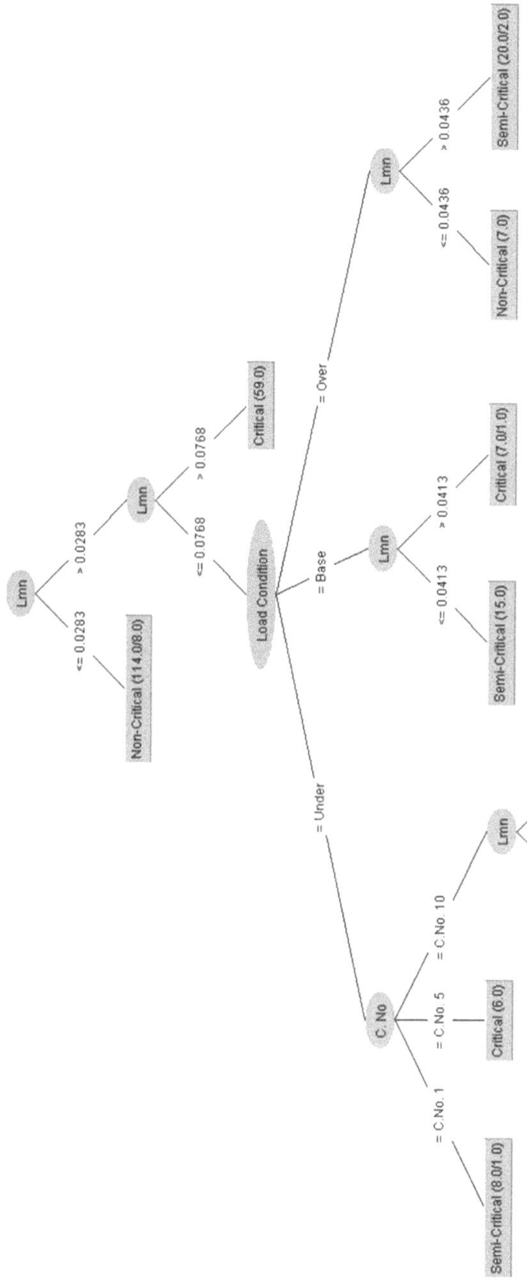

Figure 7.8 Visualizing decision tree for different load and contingencies based on Lmn.

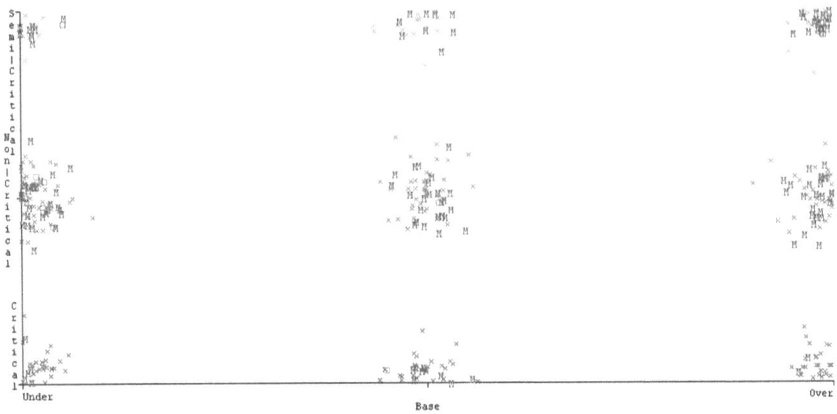

Figure 7.9 Visualizing the result of classifier error for different load condition.

Table 7.3 Trained data set for under load (70%) conditions with a variety of contingencies.

Under load condition and C. No. 1				Under load condition and C. No. 5			
Sl. No.	**L No.**	**Lmn**	**Condition**	**Sl. No.**	**L. No.**	**Lmn**	**Condition**
1.	1.	0.1118	Critical	28.	1.	0.0033	Noncritical
2.	2.	0	Noncritical	29.	2.	0.011	Noncritical
3.	3.	0.1807	Critical	30.	3.	0.0141	Noncritical
4.	4.	0.0515	Semicritical	31.	4.	0.0025	Noncritical
5.	5.	0.2966	Critical	32.	5.	0.0008	Noncritical
6.	6.	0.008	Noncritical	33.	6.	0	Noncritical
7.	7.	0.0541	Semicritical	34.	7.	0.0294	Critical
8.	8.	0.0098	Noncritical	35.	8.	0.0488	Critical
9.	9.	0.057	Semicritical	36.	9.	0.0109	Noncritical
10.	10.	0.0009	Noncritical	37.	10.	0.0083	Noncritical
11.	11.	0.2683	Critical	38.	11.	0.0347	Critical
12.	12.	0.4379	Critical	39.	12.	0.2447	Critical
13.	13.	0.0222	Noncritical	40.	13.	0.0182	Semicritical
14.	14.	0.0038	Noncritical	41.	14.	0.01	Noncritical
15.	15.	0.0748	Semicritical	42.	15.	0.0015	Noncritical
16.	16.	0.2145	Critical	43.	16.	0.0654	Critical
17.	17.	0.199	Critical	44.	17.	0.0085	Noncritical
18.	18.	0.0267	Noncritical	45.	18.	0.0162	Semicritical
19.	19.	0.0473	Semicritical	46.	19.	0.0323	Critical
20.	20.	0.2373	Critical	47.	20.	0.1121	Critical
21.	21.	0.0565	Semicritical	48.	21.	0.026	Semicritical
22.	22.	0.3606	Critical	49.	22.	0.0395	Critical
23.	23.	0.0223	Noncritical	50.	23.	0.0088	Noncritical

Table 7.3 Trained data set for under load (70%) conditions with a variety of contingencies.—cont'd

	Under load condition and C. No. 1				Under load condition and C. No. 5		
Sl. No.	**L No.**	**Lmn**	**Condition**	**Sl. No.**	**L. No.**	**Lmn**	**Condition**
24.	24.	0.0341	Noncritical	51.	24.	0.0132	Noncritical
25.	25.	0.0526	Semicritical	52.	25.	0.0176	Semicritical
26.	26.	0.0049	Noncritical	53.	26.	0.0011	Noncritical
27.	27.	0.0096	Noncritical	54.	27.	0.0032	Noncritical
	Under load condition and C. No.10				**Under load condition and C. No.10**		
Sl. No.	**L. No.**	**Lmn**	**Condition**	**Sl. No.**	**L. No.**	**Lmn**	**Condition**
55.	1.	0.0144	Noncritical	69.	15.	0.0369	Critical
56.	2.	0.0335	Critical	70.	16.	0.0071	Noncritical
57.	3.	0.0305	Semicritical	71.	17.	0.1287	Critical
58.	4.	0.0047	Noncritical	72.	18.	0.0168	Noncritical
59.	5.	0.0108	Noncritical	73.	19.	0.0324	Critical
60.	6.	0.0262	Semicritical	74.	20.	0.1189	Critical
61.	7.	0.0125	Noncritical	75.	21.	0.0286	Semicritical
62.	8.	0.0515	Critical	76.	22.	0.0362	Critical
63.	9.	0.0199	Semicritical	77.	23.	0.0085	Noncritical
64.	10.	0.0038	Noncritical	78.	24.	0.0136	Noncritical
65.	11	0.1066	Critical	79.	25.	0.0183	Semicritical
66.	12.	0.2326	Critical	80.	26.	0.0009	Noncritical
67.	13.	0.0075	Noncritical	81.	27.	0.0029	Noncritical
68.	14.	0	Noncritical	—	—	—	—

used for training to predict the severity of the line and 33% of the data is tested based on the training data set. The sample training data set for the base load condition is tabulated in Table 7.5. Based on these training data sets, severity of the other transmission line was predicted by using machine learning tool Weka, and tabulated in Table 7.6 along with various attributes.

7.3 (Case 3): Over load condition (increase of base load by 30%)

In this case, the analysis was carried out on over load condition of the system. The LVSI index was computed for base load condition, the severity was predicted based on LVSI. In this case also, 67% simulation data is used for training to predict the severity of the line and 33% of the data is tested based on the training data set. The sample training data set for the base load condition is tabulated in Table 7.7. Based on these training data sets, severity of

Table 7.4 Tested data set for under load (70%) conditions with a variety of contingencies.

Sl. No.	C. No.	L. No.	Load condition	Lmn	Predicted margin	Predicted condition
1.	'C. No. 1'	'L. No. 28'	Under	0.0369	−0.88	Semicritical
2.	'C. No. 1'	'L. No. 29'	Under	0.0033	−0.93	Noncritical
3.	'C. No. 1'	'L. No. 30'	Under	0.0144	−0.93	Noncritical
4.	'C. No. 1'	'L. No. 31'	Under	0.0326	−0.88	Semicritical
5.	'C. No. 1'	'L. No. 32'	Under	0.0196	−0.93	Noncritical
6.	'C. No. 1'	'L. No. 33'	Under	0.0823	1.00	Critical
7.	'C. No. 1'	'L. No. 34'	Under	0.0651	−0.88	Semicritical
8.	'C. No. 1'	'L. No. 35'	Under	0.0317	−0.88	Semicritical
9.	'C. No. 1'	'L. No. 36'	Under	0.0504	−0.88	Semicritical
10.	'C. No. 1'	'L. No. 37'	Under	0.0081	−0.93	Noncritical
11.	'C. No. 1'	'L. No. 38'	Under	0.0693	−0.88	Semicritical
12.	'C. No. 1'	'L. No. 39'	Under	0.0451	−0.88	Semicritical
13.	'C. No. 1'	'L. No. 40'	Under	0.0463	−0.88	Semicritical
14.	'C. No. 1'	'L. No. 41'	Under	0.002	−0.93	Noncritical
15.	'C. No. 5'	'L. No. 28'	Under	0.0114	−0.93	Noncritical
16.	'C. No. 5'	'L. No. 29'	Under	0.0006	−0.93	Noncritical
17.	'C. No. 5'	'L. No. 30'	Under	0.0027	−0.93	Noncritical
18.	'C. No. 5'	'L. No. 31'	Under	0.0205	−0.93	Noncritical
19.	'C. No. 5'	'L. No. 32'	Under	0.0155	−0.93	Noncritical
20.	'C. No. 5'	'L. No. 33'	Under	0.0296	1.00	Critical
21.	'C. No. 5'	'L. No. 34'	Under	0.0106	−0.93	Noncritical
22.	'C. No. 5'	'L. No. 35'	Under	0.019	−0.93	Noncritical
23.	'C. No. 5'	'L. No. 36'	Under	0.0252	−0.93	Noncritical
24.	'C. No. 5'	'L. No. 37'	Under	0.0044	−0.93	Noncritical
25.	'C. No. 5'	'L. No. 38'	Under	0.0317	1.00	Critical

26.	'C. No. 5'	'L. No. 39'	Under	0.0218	−0.93	Noncritical
27.	'C. No. 5'	'L. No. 40'	Under	0.0174	−0.93	Noncritical
28.	'C. No. 5'	'L. No. 41'	Under	0.001	−0.93	Noncritical
29.	'C. No. 10'	'L. No. 28'	Under	0.0119	−0.93	Noncritical
30.	'C. No. 10'	'L. No. 29'	Under	0.0006	−0.93	Noncritical
31.	'C. No. 10'	'L. No. 30'	Under	0.0018	−0.93	Noncritical
32.	'C. No. 10'	'L. No. 31'	Under	0.0212	−0.93	Noncritical
33.	'C. No. 10'	'L. No. 32'	Under	0.0106	−0.93	Noncritical
34.	'C. No. 10'	'L. No. 33'	Under	0.0404	1.00	Critical
35.	'C. No. 10'	'L. No. 34'	Under	0.0115	−0.93	Noncritical
36.	'C. No. 10'	'L. No. 35'	Under	0.0256	−0.93	Noncritical
37.	'C. No. 10'	'L. No. 36'	Under	0.0252	−0.93	Noncritical
38.	'C. No. 10'	'L. No. 37'	Under	0.0167	−0.93	Noncritical
39.	'C. No. 10'	'L. No. 38'	Under	0.0315	1.00	Critical
40.	'C. No. 10'	'L. No. 39'	Under	0.0217	−0.93	Noncritical
41.	'C. No. 10'	'L. No. 40'	Under	0.0122	−0.93	Noncritical
42.	'C. No. 10'	'L. No. 41'	Under	0.001	−0.93	Noncritical

Table 7.5 Trained data set for base load (100%) conditions with a variety of contingencies.

	Base load condition and C. No.1				Base load condition and C. No.5		
Sl. No.	**L. No.**	**Lmn**	**Condition**	**Sl. No.**	**L. No.**	**Lmn**	**Condition**
1.	1.	0.1015	Critical	28.	1.	0.0760	Critical
2.	2.	0.0000	Noncritical	29.	2.	0.0245	Noncritical
3.	3.	0.0853	Critical	30.	3.	0.0017	Noncritical
4.	4.	0.0220	Noncritical	31.	4.	0.0187	Noncritical
5.	5.	0.1480	Critical	32.	5.	0.0000	Noncritical
6.	6.	0.0034	Noncritical	33.	6.	0.0610	Critical
7.	7.	0.0341	Semicritical	34.	7.	0.0452	Semicritical
8.	8.	0.0149	Noncritical	35.	8.	0.0261	Noncritical
9.	9.	0.0332	Semicritical	36.	9.	0.1108	Critical
10.	10.	0.0037	Noncritical	37.	10.	0.0312	Semicritical
11.	11.	0.0902	Critical	38.	11.	0.0541	Critical
12.	12.	0.3223	Critical	39.	12.	0.3074	Critical
13.	13.	0.0202	Noncritical	40.	13.	0.0172	Noncritical
14.	14.	0.0069	Noncritical	41.	14.	0.0105	Noncritical
15.	15.	0.0314	Semicritical	42.	15.	0.0246	Noncritical
16.	16.	0.1227	Critical	43.	16.	0.1196	Critical
17.	17.	0.0833	Critical	44.	17.	0.0874	Critical
18.	18.	0.0187	Noncritical	45.	18.	0.0206	Noncritical
19.	19.	0.0333	Semicritical	46.	19.	0.0377	Semicritical
20.	20.	0.1683	Critical	47.	20.	0.1686	Critical
21.	21.	0.0395	Semicritical	48.	21.	0.0398	Semicritical
22.	22.	0.2132	Critical	49.	22.	0.1771	Critical
23.	23.	0.0162	Noncritical	50.	23.	0.0148	Noncritical
24.	24.	0.0249	Noncritical	51.	24.	0.0234	Noncritical
25.	25.	0.0403	Semicritical	52.	25.	0.0360	Semicritical
26.	26.	0.0036	Noncritical	53.	26.	0.0027	Noncritical
27.	27.	0.0076	Noncritical	54.	27.	0.0064	Noncritical
	Base load condition and C. No.10				**Base load condition and C. No.10**		
Sl. No.	**L. No.**	**Lmn**	**Condition**	**Sl. No.**	**L. No.**	**Lmn**	**Condition**
55.	1.	0.0372	Semicritical	69.	15.	0.0432	Critical
56.	2.	0.0151	Noncritical	70.	16.	0.0096	Noncritical
57.	3.	0.0170	Noncritical	71.	17.	0.1798	Critical
58.	4.	0.0113	Noncritical	72.	18.	0.0161	Noncritical
59.	5.	0.0314	Semicritical	73.	19.	0.0246	Noncritical
60.	6.	0.0064	Noncritical	74.	20.	0.1638	Critical
61.	7.	0.0297	Semicritical	75.	21.	0.0388	Semicritical
62.	8.	0.0413	Semicritical	76.	22.	0.1861	Critical
63.	9.	0.0442	Critical	77.	23.	0.0161	Noncritical

Table 7.5 Trained data set for base load (100%) conditions with a variety of contingencies.—cont'd

	Base load condition and C. No.10				Base load condition and C. No.10		
Sl. No.	**L. No.**	**Lmn**	**Condition**	**Sl. No.**	**L. No.**	**Lmn**	**Condition**
64.	10.	0.0019	Noncritical	78.	24.	0.0277	Semicritical
65.	11.	0.0111	Noncritical	79.	25.	0.0520	Critical
66.	12.	0.3236	Critical	80.	26.	0.0034	Noncritical
67.	13.	0.0170	Noncritical	81.	27.	0.0092	Noncritical
68.	14.	0.0000	Noncritical	–	–	–	–

the other transmission line was predicted by using machine learning tool Weka, and tabulated in Table 7.8 along with various attributes.

The detailed summary of the simulation as follows: time taken to test model on training data: 0.01 s, correctly classified instances are 231(95.0617%), incorrectly classified instances are 12 (4.9383%). Also, it observed performance measures such as Kappa statistics, Mean absolute error, Root mean squared error, relative absolute squared error, and root relative squared error were analyzed; all the values are found within the permissible limits. Based on the training data set, the sample missing data set condition is predicted for the under load, base load, and over load conditions under single line outage as mentioned in Tables 7.4, 7.6 and 7.8 respectively, and the proposed method was found to be effective when manually verified.

8. Conclusion

This study employs a role of Big Data and machine learning approach to identify and predict the severity of the transmission line. In the Big Data analytics, data mining has been considered for preprocessing the data obtained for contingency. Classification algorithm is applied to identify and predict the severity. The line voltage stability index was computed to rank the severity of the lines. It had led to a huge volume of data for the contingency analysis. Sixty-six percentage of the data was used to train the system, and remaining 33% of data was tested. The severity of the line was categorized into Critical, Semicritical, and Noncritical based on classification algorithm technique. The proposed method was found to be effective and was able to obtain the mean absolute error close to 0.0605% compared to manual categorization.

Table 7.6 Tested data set for base load (100%) conditions with a variety of contingencies.

Sl. No.	C. No.	L. No.	Load condition	Lmn	Predicted margin	Predicted condition
1.	'C. No. 1'	'L. No. 28'	Base	0.0262	−0.93	Noncritical
2.	'C. No. 1'	'L. No. 29'	Base	0.0026	−0.93	Noncritical
3.	'C. No. 1'	'L. No. 30'	Base	0.0120	−0.93	Noncritical
4.	'C. No. 1'	'L. No. 31'	Base	0.0223	−0.93	Noncritical
5.	'C. No. 1'	'L. No. 32'	Base	0.0194	−0.93	Noncritical
6.	'C. No. 1'	'L. No. 33'	Base	0.0502	0.71	Critical
7.	'C. No. 1'	'L. No. 34'	Base	0.0446	0.71	Critical
8.	'C. No. 1'	'L. No. 35'	Base	0.0263	−0.93	Noncritical
9.	'C. No. 1'	'L. No. 36'	Base	0.0370	−1.00	Semicritical
10.	'C. No. 1'	'L. No. 37'	Base	0.0017	−0.93	Noncritical
11.	'C. No. 1'	'L. No. 38'	Base	0.0484	0.71	Critical
12.	'C. No. 1'	'L. No. 39'	Base	0.0325	−1.00	Semicritical
13.	'C. No. 1'	'L. No. 40'	Base	0.0291	−1.00	Semicritical
14.	'C. No. 1'	'L. No. 41'	Base	0.0014	−0.93	Noncritical
15.	'C. No. 5'	'L. No. 28'	Base	0.0232	−0.93	Noncritical
16.	'C. No. 5'	'L. No. 29'	Base	0.0021	−0.93	Noncritical
17.	'C. No. 5'	'L. No. 30'	Base	0.0088	−0.93	Noncritical
18.	'C. No. 5'	'L. No. 31'	Base	0.0254	−0.93	Noncritical
19.	'C. No. 5'	'L. No. 32'	Base	0.0191	−0.93	Noncritical
20.	'C. No. 5'	'L. No. 33'	Base	0.0512	0.71	Critical
21.	'C. No. 5'	'L. No. 34'	Base	0.0350	−1.00	Semicritical
22.	'C. No. 5'	'L. No. 35'	Base	0.0225	−0.93	Noncritical
23.	'C. No. 5'	'L. No. 36'	Base	0.0370	−1.00	Semicritical
24.	'C. No. 5'	'L. No. 37'	Base	0.0059	−0.93	Noncritical
25.	'C. No. 5'	'L. No. 38'	Base	0.0484	0.71	Critical

26.	‘C. No. 5’	‘L. No. 39’	Base	0.0325	−1.00	Semicritical
27.	‘C. No. 5’	‘L. No. 40’	Base	0.0322	−1.00	Semicritical
28.	‘C. No. 5’	‘L. No. 41’	Base	0.0014	−0.93	Noncritical
29.	‘C. No. 10’	‘L. No. 28’	Base	0.0268	−0.93	Noncritical
30.	‘C. No. 10’	‘L. No. 29’	Base	0.0037	−0.93	Noncritical
31.	‘C. No. 10’	‘L. No. 30’	Base	0.0159	−0.93	Noncritical
32.	‘C. No. 10’	‘L. No. 31’	Base	0.0182	−0.93	Noncritical
33.	‘C. No. 10’	‘L. No. 32′	Base	0.0237	−0.93	Noncritical
34.	‘C. No. 10’	‘L. No. 33’	Base	0.0429	0.71	Critical
35.	‘C. No. 10’	‘L. No. 34’	Base	0.0445	0.71	Critical
36.	‘C. No. 10’	‘L. No. 35’	Base	0.0208	−0.93	Noncritical
37.	‘C. No. 10’	‘L. No. 36’	Base	0.0367	−1.00	Semicritical
38.	‘C. No. 10’	‘L. No. 37’	Base	0.0120	−0.93	Noncritical
39.	‘C. No. 10’	‘L. No. 38’	Base	0.0479	0.71	Critical
40.	‘C. No. 10’	‘L. No. 39’	Base	0.0322	−1.00	Semicritical
41.	‘C. No. 10’	‘L. No. 40’	Base	0.0344	−1.00	Semicritical
42.	‘C. No. 10’	‘L. No. 41’	Base	0.0014	−0.93	Noncritical

Table 7.7 Trained data set for over load (130%) conditions with a variety of contingencies.

	Base load condition and C. No. 1				Base load condition and C. No. 5		
Sl. No.	**L. No.**	**Lmn**	**Condition**	**Sl. No.**	**L. No.**	**Lmn**	**Condition**
1.	1.	0.1118	Critical	28.	1.	0.0230	Noncritical
2.	2.	0.0000	Noncritical	29.	2.	0.0323	Noncritical
3.	3.	0.1807	Critical	30.	3.	0.0768	Semicritical
4.	4.	0.0515	Semicritical	31.	4.	0.0503	Semicritical
5.	5.	0.2966	Critical	32.	5.	0.0000	Noncritical
6.	6.	0.0080	Noncritical	33.	6.	0.1174	Critical
7.	7.	0.0541	Semicritical	34.	7.	0.0862	Critical
8.	8.	0.0098	Noncritical	35.	8.	0.0006	Noncritical
9.	9.	0.0570	Semicritical	36.	9.	0.2590	Critical
10.	10.	0.0009	Noncritical	37.	10.	0.0870	Critical
11.	11.	0.2683	Critical	38.	11.	0.2391	Critical
12.	12.	0.4379	Critical	39.	12.	0.4192	Critical
13.	13.	0.0222	Noncritical	40.	13.	0.0177	Noncritical
14.	14.	0.0038	Noncritical	41.	14.	0.0068	Noncritical
15.	15.	0.0748	Semicritical	42.	15.	0.0694	Semicritical
16.	16.	0.2145	Critical	43.	16.	0.2116	Critical
17.	17.	0.1990	Critical	44.	17.	0.2081	Critical
18.	18.	0.0267	Noncritical	45.	18.	0.0302	Noncritical
19.	19.	0.0473	Semicritical	46.	19.	0.0555	Semicritical
20.	20.	0.2373	Critical	47.	20.	0.2386	Critical
21.	21.	0.0565	Semicritical	48.	21.	0.0573	Semicritical
22.	22.	0.3606	Critical	49.	22.	0.3211	Critical
23.	23.	0.0223	Noncritical	50.	23.	0.0201	Noncritical
24.	24.	0.0341	Noncritical	51.	24.	0.0315	Noncritical
25.	25.	0.0526	Semicritical	52.	25.	0.0446	Noncritical
26.	26.	0.0049	Noncritical	53.	26.	0.0036	Noncritical
27.	27.	0.0096	Noncritical	54.	27.	0.0076	Noncritical

	Base load condition and C. No. 10				Base load condition and C. No. 10		
Sl. No.	**L. No.**	**Lmn**	**Condition**	**Sl. No.**	**L. No.**	**Lmn**	**Condition**
55.	1.	0.0261	Noncritical	69.	15.	0.0672	Critical
56.	2.	0.0460	Semicritical	70.	16.	0.0283	Noncritical
57.	3.	0.0309	Noncritical	71.	17.	0.3080	Critical
58.	4.	0.0149	Noncritical	72.	18.	0.0171	Noncritical
59.	5.	0.0437	Semicritical	73.	19.	0.0204	Noncritical
60.	6.	0.0026	Noncritical	74.	20.	0.2176	Critical
61.	7.	0.0428	Noncritical	75.	21.	0.0512	Semicritical
62.	8.	0.0436	Noncritical	76.	22.	0.3400	Critical
63.	9.	0.0519	Semicritical	77.	23.	0.0250	Noncritical
64.	10.	0.0014	Noncritical	78.	24.	0.0437	Semicritical
65.	11.	0.1252	Critical	79.	25.	0.0891	Critical
66.	12.	0.5461	Critical	80.	26.	0.0061	Noncritical
67.	13.	0.0473	Semicritical	81.	27.	0.0160	Noncritical
68.	14.	0.0000	Noncritical	—	—	—	—

Table 7.8 Tested data set for over load (130%) conditions with a variety of contingencies.

Sl. No.	C. No.	L. No.	Load condition	Lmn	Predicted margin	Predicted condition
1.	'C. No. 1'	'L. No. 28'	Over	0.0369	−1.00	Noncritical
2.	'C. No. 1'	'L. No. 29'	Over	0.0033	−0.93	Noncritical
3.	'C. No. 1'	'L. No. 30'	Over	0.0144	−0.93	Noncritical
4.	'C. No. 1'	'L. No. 31'	Over	0.0326	−1.00	Noncritical
5.	'C. No. 1'	'L. No. 32'	Over	0.0196	−0.93	Noncritical
6.	'C. No. 1'	'L. No. 33'	Over	0.0823	1.00	Critical
7.	'C. No. 1'	'L. No. 34'	Over	0.0651	−0.85	Semicritical
8.	'C. No. 1'	'L. No. 35'	Over	0.0317	−1.00	Noncritical
9.	'C. No. 1'	'L. No. 36'	Over	0.0504	−0.85	Semicritical
10.	'C. No. 1'	'L. No. 37'	Over	0.0081	−0.93	Noncritical
11.	'C. No. 1'	'L. No. 38'	Over	0.0693	−0.85	Semicritical
12.	'C. No. 1'	'L. No. 39'	Over	0.0451	−0.85	Semicritical
13.	'C. No. 1'	'L. No. 40'	Over	0.0463	−0.85	Semicritical
14.	'C. No. 1'	'L. No. 41'	Over	0.0020	−0.93	Noncritical
15.	'C. No. 5'	'L. No. 28'	Over	0.0318	−1.00	Noncritical
16.	'C. No. 5'	'L. No. 29'	Over	0.0023	−0.93	Noncritical
17.	'C. No. 5'	'L. No. 30'	Over	0.0087	−0.93	Noncritical
18.	'C. No. 5'	'L. No. 31'	Over	0.0381	−1.00	Noncritical
19.	'C. No. 5'	'L. No. 32'	Over	0.0188	−0.93	Noncritical
20.	'C. No. 5'	'L. No. 33'	Over	0.0852	1.00	Critical
21.	'C. No. 5'	'L. No. 34'	Over	0.0489	−0.85	Semicritical
22.	'C. No. 5'	'L. No. 35'	Over	0.0259	−0.93	Noncritical
23.	'C. No. 5'	'L. No. 36'	Over	0.0502	−0.85	Semicritical
24.	'C. No. 5'	'L. No. 37'	Over	0.0197	−0.93	Noncritical
25.	'C. No. 5'	'L. No. 38'	Over	0.0694	−0.85	Semicritical

(Continued)

Table 7.8 Tested data set for over load (130%) conditions with a variety of contingencies.—cont'd

Sl. No.	C. No.	L. No.	Load condition	Lmn	Predicted margin	Predicted condition
26.	'C. No. 5'	'L. No. 39'	Over	0.0451	−0.85	Semicritical
27.	'C. No. 5'	'L. No. 40'	Over	0.0508	−0.85	Semicritical
28.	'C. No. 5'	'L. No. 41'	Over	0.0020	−0.93	Noncritical
29.	'C. No. 10'	'L. No. 28'	Over	0.0427	−1.00	Noncritical
30.	'C. No. 10'	'L. No. 29'	Over	0.0069	−0.93	Noncritical
31.	'C. No. 10'	'L. No. 30'	Over	0.0297	−1.00	Noncritical
32.	'C. No. 10'	'L. No. 31'	Over	0.0175	−0.93	Noncritical
33.	'C. No. 10'	'L. No. 32'	Over	0.0393	−1.00	Noncritical
34.	'C. No. 10'	'L. No. 33'	Over	0.0467	−0.85	Semicritical
35.	'C. No. 10'	'L. No. 34'	Over	0.0767	−0.85	Semicritical
36.	'C. No. 10'	'L. No. 35'	Over	0.0126	−0.93	Noncritical
37.	'C. No. 10'	'L. No. 36'	Over	0.0500	−0.85	Semicritical
38.	'C. No. 10'	'L. No. 37'	Over	0.0544	−0.85	Semicritical
39.	'C. No. 10'	'L. No. 38'	Over	0.0691	−0.85	Semicritical
40.	'C. No. 10'	'L. No. 39'	Over	0.0450	−0.85	Semicritical
41.	'C. No. 10'	'L. No. 40'	Over	0.0641	−0.85	Semicritical
42.	'C. No. 10'	'L. No. 41'	Over	0.0019	−0.93	Noncritical

Acknowledgments

The authors like to express their gratitude to the managements of Presidency University, Bengaluru, and Sree Vidyanikethan Engineering College, Tirupati, for their continual support and provision of facilities necessary for the completion of this study.

References

1. Akhavan-Hejazi H, Mohsenian-Rad H. Power systems big data analytics: an assessment of paradigm shift barriers and prospects. *Energy Rep*. 2018;4:91−100. https://doi.org/10.1016/j.egyr.2017.11.002.
2. Morais J, Pires Y, Cardoso C, Klautau A. *An Overview of Data Mining Techniques Applied to Power Systems*. 2009, 10.5772/6463.
3. Angadi RV, Babu Daram S, Venkataramu PS. Analysis of power system security using big data and machine learning techniques. In: *2020 IEEE 17th India Council International Conference, INDICON 2020*. 2020. https://doi.org/10.1109/INDICON49873.2020.9342458.
4. Angadi RV, Babu Daram S, Pravallika D, Venkataramu PS. Contingency ranking through big data analytics with GIPFC in power system under n-1condition. In: *2020 IEEE 17th India Council International Conference, INDICON 2020*. 2020. https://doi.org/10.1109/INDICON49873.2020.9342527.
5. Angadi RV, Daram SB, Venkataramu PS. Contingency analysis of power system using big data analytic techniques. In: *Proceedings of the 2020 International Conference on Computing, Communication and Security, ICCCS 2020*. 2020. https://doi.org/10.1109/ICCCS49678.2020.9276796.
6. Angadi RV, Venkataramu PS, Daram SB. Role of big data analytics in power system Application. In: *E3S Web of Conferences*. 2020:184. https://doi.org/10.1051/e3sconf/202018401017.
7. Angadi VS, Daram B. Severity prediction of single transmission line outage using big data and machine learning techniques. *Int J Grid Distrib Comput*. 2020;13:1090−1108.
8. Boicea VA. *Energy Management: Big Data in Power Load Forecasting*. 2021. https://doi.org/10.1201/9781003147213.
9. Bulat H, Franković D, Vlahinić S. Enhanced contingency analysis—a power system operator tool. *Energies*. 2021;14(4):923. https://doi.org/10.3390/en14040923.
10. Chu CC, Lee SH, Chuang HY. Efficient look-ahead load margin and voltage profiles contingency analysis using the tangent vector index method. In: *Proceedings of the IEEE Power Engineering Society Transmission and Distribution Conference*. 1. 2000:225−230.
11. Daram SB, Venkataramu PS, Nagaraj MS. Performance index based contingency ranking under line outage condition incorporating IPFC. In: *International Conference on Electrical, Electronics, and Optimization Techniques, ICEEOT 2016*. 2016: 2589−2593. https://doi.org/10.1109/ICEEOT.2016.7755161.
12. Daram V, Nagaraj MS. Condition number based contingency ranking under line outage condition incorporating IPFC. In: *International Conference on Renewable Energy Utilization*. 2016.
13. Daram SB, Venkataramu PS, Nagaraj MS. Voltage stability index based critical bus ranking incorporating IPFC under single transmission line outage condition. *Int J Electr Power Syst Technol*. 2016;2:12−19.
14. Data mining technology and its applications to power systems. *Int J Comput Appl*. 2015; 131(8):1−6. https://doi.org/10.5120/ijca2015907273.
15. Deeba K, Amutha B. Classification algorithms of data mining. *Indian J Sci Technol*. 2016; 9(39). https://doi.org/10.17485/ijst/2016/v9i39/102065.

16. García S, Luengo J, Herrera F. *Data Preprocessing in Data Mining*. Vol. 72. Intelligent Systems Reference Library; 2015. http://www.springer.com/series/8578?detailsPage=titles.
17. García S, Ramírez-Gallego S, Luengo J, Benítez JM, Herrera F. Big data preprocessing: methods and prospects. *Big Data Analytics*. 2016. https://doi.org/10.1186/s41044-016-0014-0.
18. Gjorgiev B, Sansavini G. *Identifying and Assessing Power System Vulnerabilities to Transmission Asset Outages via Cascading Failure Analysis*. Reliability Engineering & System Safety; 2021:108085. https://doi.org/10.1016/j.ress.2021.108085.
19. Guo Y, Yang Z, Feng S, Hu J. Complex power system status monitoring and evaluation using big data platform and machine learning algorithms: a review and a case study. *Complexity*. 2018. https://doi.org/10.1155/2018/8496187. Article ID.8496187.
20. Halim RA, Gabash A, Li P. Sensitivity analysis of power flow Jacbian matrix in active distribution networks. *IEEE Power Energy Student Summit (PESS)*. 2014:49–54.
21. Miraftabzadeh SM, Foiadelli F, Longo M, Pasetti M. A survey of machine learning applications for power system Analytics. In: *2019 IEEE International Conference on Environment and Electrical Engineering*. 2019:1–5. https://doi.org/10.1109/EEEIC.2019.8783340.
22. Mun S-D. A study on the development of automation algorithm for refining asset data of overhead transmission lines. *KEPCO J Electr Power Energy*. 2021;7:73–77. https://doi.org/10.18770/KEPCO.2021.07.01.073.
23. Pires Y, Morais J, Cardoso C, Klautau A. Data mining applied to the electric power industry: classification of short-circuit faults in transmission lines. *Stud Comput Intell*. 2009;169:107–122. https://doi.org/10.1007/978-3-540-88045-5_6.
24. Rani K, Amarnath J, Kamakshaiah S. Contingency analysis under deregulated power systems. *ICGST ACSE J*. 2011;11.
25. Ridzuan F, Wan ZWMN. A review on data cleansing methods for big data. *Procedia Comput Sci*. 2019;161:731–738. https://doi.org/10.1016/j.procs.2019.11.177.
26. Shafaat SA, Teja PV, Reddy PSK, Manoj R, Daram SB, Obulesu O. Application of big data analytics in power system under single transmission line outage condition. In: *Proceedings of the International Conference on Trends in Electronics and Informatics, ICOEI 2019*. 2019:377–382. https://doi.org/10.1109/ICOEI.2019.8862685.
27. Witten IH, Frank E, Hall MA. *Data Mining Practical Machine Learning Tools and Techniques*. 3rd ed.
28. Zheng J, Dagnino A. An initial study of predictive machine learning analytics on large volumes of historical data for power system applications. In: *Proceedings—2014 IEEE International Conference on Big Data, IEEE Big Data 2014*. 2014:952–959. https://doi.org/10.1109/BigData.2014.7004327.

Further reading

1. Jain M, Venkataramu PS, Ananthapadmanabha T. Critical bus ranking under line outage contingencies. In: *Proceedings of the IASTED International Conference on Energy and Power Systems*. 2007:69–74.
2. Kothari D.P., Nagrath I.J., Modern Power System Analysis. In: McGraw Hill Edition (India) Private Limited.

CHAPTER EIGHT

Ensemble classifier-based protection scheme for hybrid microgrid

Goyal Awagan, Ebha Koley and Subhojit Ghosh
Department of Electrical Engineering, National Institute of Technology, Raipur, Chhattisgarh, India

Contents

1. Introduction

The concern for greenhouse gas emission, rising energy demand, and technological advancements has driven the focus of consumers and power utilities toward the integration of DERs in the centralized structure of power system.[1] This centralized structure based on conventional hierarchy allows unidirectional power flow from the transmission level to the distribution level. However, this classical structure has certain limitations due lesser penetration of DER units, high power losses, poor power quality, low resiliency to faults, and reduced reliability. These limitations of

Smart Electrical and Mechanical Systems
ISBN: 978-0-323-90789-7
https://doi.org/10.1016/B978-0-323-90789-7.00003-8

conventional power system are addressed using the microgrid concept which allows for the proliferation of DERs with improved grid resiliency.[2,3]

Microgrids are categorized as AC, DC, and hybrid depending upon the nature of the voltage waveforms and current waveforms obtained at each node in the network. The simple architecture, ease of modeling, quick transformation at varying voltages, and flexibility of integration of DER units have attracted the attention of AC microgrids over the past decades. However, AC microgrid demands proper synchronization of DERs with the grid and power conditioning unit(s) consisting of DC/AC converters. The limitations imposed by AC microgrids have attracted the attention of DC microgrids which allows for direct integration of loads and DERs powered by DC. The DC microgrids have added to higher transmission capacity and improved efficiency due to reduced number of interfacing devices. However, having pure DC microgrid is not realistic as it needs complete modifications of the existing system. With the concept of hybrid microgrid, the limitations offered by AC as well as DC microgrids are nullified by their individual advantages.[4,5]

The rising penetration of DERs in the present system is imposing certain technical challenges in the development of a reliable, robust, and economical microgrid protection scheme.[6] The protection is further complicated due to the intermittent nature of DERs,[7] faults in DERs and early islanding. A number of protection schemes addressing the issues due to the variations in short-circuit current levels in microgrid operating modes, that is, grid-connected and islanding,[8] intermittency in the nature of wind and solar-based DERs[9] have been reported in the literature for AC microgrids. Several protection schemes have also been reported for DC microgrids.[10] However, the volume of reported literature on the challenges in the protection of hybrid microgrid is quite limited. A brief review discussing the challenges in hybrid microgrid protection and future directions has been carried out in.[11,12] Fundamentally, the protection of hybrid microgrid is in itself a very challenging task as both AC and DC generation are involved. Schemes proposed for AC microgrids cannot be directly adapted for hybrid systems because current signals at the DC buses do not cross zero. Similarly, because of the bipolar nature of the current waveform and voltage waveform at the AC buses, schemes proposed for DC microgrids cannot be directly replicated for Hybrid networks. The scheme proposed in the present work is hybrid in nature since it combines both the AC and DC protection scheme.

Considering all the above challenges, a widely used machine-learning approach using ensemble subspace discriminant (ESD) classifiers is designed for a hybrid microgrid system which can detect and distinguish between all types of line faults and perform the faulty section identification. The proposed protection scheme is hybrid in nature which combines both the AC and DC protection scheme. In a power system, sensors are usually installed to accurately monitor the real-time voltage and/or current phasors of each bus. The voltage and/or current measurements derived from each bus are useful for knowing the state of the system. The states estimated from the sensor measurements are used to carry out the stability and vulnerability analysis. However, state estimation should be accomplished with minimal sensor set such that the observability of the network is attained with minimum measurements. While framing the problem of optimal sensor placement, the cost of sensors and the associated communication infrastructure should be taken into consideration.[13,14] In the present work, optimal sensor placement is framed around the objective of choosing a minimum number of buses for allocating the sensors for full system observability and solved using genetic algorithm (GA).

The proposed hybrid microgrid has been designed utilizing discrete wavelet transforms (DWT), which use the multiresolution technique to derive discriminatory features. The signal analysis is carried out in DWT by splitting the input signal into different frequency ranges.[15] The original signal is decomposed into approximate coefficients consisting of information in the low-frequency range and detailed coefficients consisting of information in the high-frequency ranges. For the detection and characterization of faults as well as the identification of faulty lines, approximate coefficients of low frequency are used.[16] The ESD classifier approach is adopted for performing the classification tasks. This ensemble-based classifier outperforms the single classifier by achieving high classification accuracy and improved resiliency against noninformative features.[17] Each classifier module is responsible for detecting/classifying faults and identifying faulty sections simultaneously. The classifier intended for section identification identifies the faulty section and generates a trip signal for its isolation. With prompt isolation of the faulty section, the reliability of power supply is maintained as the loads are supplied by the healthy section and the entire system does not require total shutdown.

The significant contributions of the work proposed in this chapter are outlined as:

1. Design of a hybrid microgrid protection scheme which utilizes DWT and ESD, with separate modules for detecting/classifying faults in different sections of the microgrid for both the operating modes.

2. Optimization-based approach has been proposed for determining strategic sensor locations in the hybrid network based on observability criteria.
3. The information from identified sensors has been utilized to formulate an ensemble-based classifier for protection of hybrid network.

Here is how the rest of the chapter is structured: Section 2 illustrates the schematic and modeling of the hybrid microgrid system under study. Section 3 discusses the GA-based optimal sensor location problem. Using an ensemble classifier, the hybrid microgrid protection scheme is developed in Section 4 and its performance is evaluated in Section 5. The chapter concludes with Section 6.

2. Modeling and simulation of hybrid microgrid system

The hybrid microgrid under study is schematically depicted in Fig. 8.1.[18] The hybrid microgrid is connected at 69 kV, 60 Hz to the

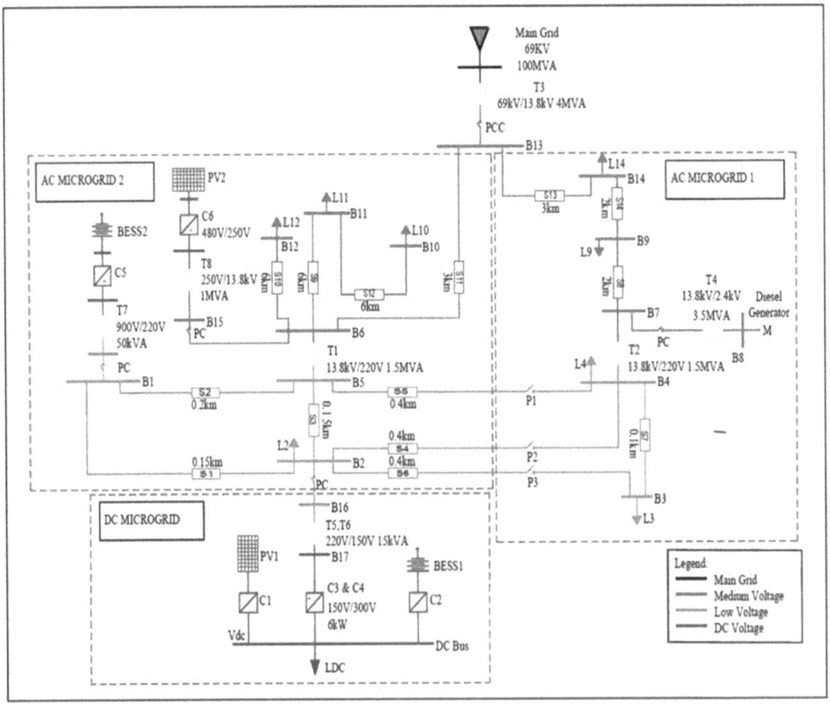

Figure 8.1 Hybrid microgrid model under study. From Orti L, Orizondo R, Águila A, González, JW, López, GJ, Isaac, I. Hybrid AC/DC microgrid test system simulation: grid-connected mode. *Heliyon*. 2019;5(12). https://doi.org/10.1016/j.heliyon.2019.e02862.

main grid and operates at voltage levels 13.8 kV (primary voltage) and 0.22 kV (secondary voltage). This microgrid consists of five DER units, that is, one diesel generator of 3 MW, two photovoltaic distributed generators (PVDG), that is, PVDG1 of 10.5 kW, PVDG2 of 725 kW and two battery energy storage systems (BESS), that is, BESS1 operating with a single battery unit of 800 Ah at 120 V DC and BESS2 operating with three battery units of 1.5 Ah at 650 V DC. This microgrid has three submicrogrids, namely, AC microgrid 1, AC microgrid 2 and DC microgrid. PVDG1 and BESS1 are integrated with a DC–DC boost converter (C1) and buck-boost bidirectional converter (C2), respectively, in the DC microgrid area. DC busbar is connected to the AC microgrid 2 area through two DC-AC bidirectional converters (C3 & C4) connected in parallel through transformers T5 and T6, respectively. Similarly, BESS2 and PVDG2 operate with a DC-AC bidirectional converter (C5) through transformer T7 and DC-AC PWM inverter (C6) through transformer T8 respectively in AC microgrid 2. The distribution network has 14 buses numbered as B1, B2, B3, …….B14 and 14 lines, namely, S_1, S_2, S_3, …….S_{14}. L3, L4, L9, L10, L11, L12, L14 and LDC are loads at buses B3, B4, B9, B10, B11, B12, B14 and DC bus, respectively.

3. Selection of critical buses for sensor placement using optimization algorithm

The sensor installed at any particular bus is able to provide the necessary information regarding the voltage-current dynamics of all the lines connected to that bus. If sensor installed at any particular bus location provides the necessary information regarding the system behavior, the corresponding bus and its adjoining busses are said to be observable.[19] According to K. A. Clements,[20] there are two concepts for assessing the observability of the linear system. The first concept is based on the topology of the network (topological observability) and second concept is based on the location and types of measurement (numerical observability). The task of optimal sensor placement determines the minimum set of sensors that needs to be installed for the complete system to be observable. In an observable system, each bus is observable by a subset of the entire set of sensors. A system in which a particular bus is found to be observable more than once provides more robustness and enhanced reliability.[21]

For n-bus hybrid microgrid system, the generalized problem for optimal sensor placement can be formulated as follows[22]:

$$\min \sum_{i=1}^{n} w_i x_i \text{s.t. } T_S X \geq BX = [x_1 \; x_2 \; \dots \dots \dots \; x_n]^T x_i \in \{0, 1\} \quad (8.1)$$

where, w_i is the cost of sensor placed at i^{th} bus.

x_i is the sensor placement variable,

$$B = [1\ 1\ \dots\ \dots\ \dots]^T_{n\times 1}$$

T_S the connectivity matrix whose entries are,

$$t_{k,l} = \begin{cases} 1, & \text{if } k = l \\ 1, & \text{if } k \text{ and } l \text{ are connected} \\ 0, & \text{otherwise} \end{cases} \tag{8.2}$$

The optimal sensor placement problem is framed using the metaheuristic GA approach which relies on the principles of natural selection and biological evolution. Conventional optimization techniques have limitations for finding the global optimal solution and are also trapped in local optima. However, GA begins with the random generation of initial population which guarantees the global optimum solution to be reached. The new population (children) is produced using the random individuals from the initial population (parents). The GA operators, that is, selection, crossover, and mutation, are applied to the parent members during each generation to find the solution with the maximum fitness.[23,24]

The fitness function (z) for optimal sensor placement is defined with an objective to obtain the smallest value of z satisfying the requirements of good quality sensor allocation plan. An initial population of 128 chromosomes is chosen randomly. As mentioned earlier in ILP, there are 17 variables corresponding to each bus in the system under study where each variable can either take value one or zero depending on the sensor placed at that bus. Thus, each chromosome consists of 17 randomly chosen bits. Crossovers were allowed at a rate of 0.5 and mutations were allowed at a rate of 0.1. The flowchart for GA is shown in Fig. 8.2. The lower and the upper bounds on the variables were set as 0 and 1, respectively. The optimal solution was obtained using Roulette selection, single-point crossover, and adaptive feasible mutation parameters for 56 iterations. The optimal solution obtained for sensor placement problem using GA is found in MATLAB which indicates that the sensors must be installed at six bus locations, that is, at buses 2, 6, 8, 9, 10, and 17 in order to make the whole microgrid system observable.

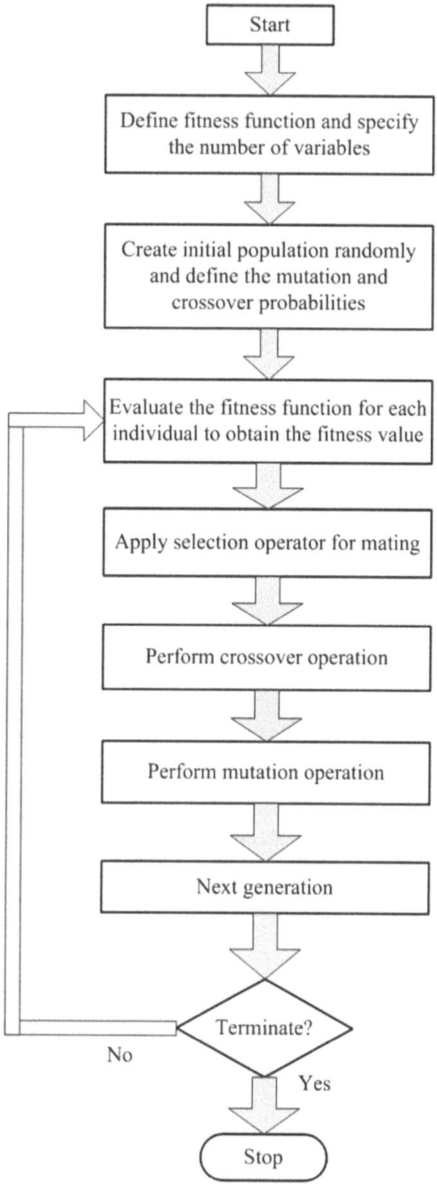

Figure 8.2 Flowchart of genetic algorithm.

4. Development of ESD-based protection scheme for hybrid microgrid

The development of an overall protection scheme for hybrid microgrid system using the ESD approach is discussed in this section for both the microgrid operating modes. The protection scheme is designed using the voltage and current data obtained from the optimal number of buses as mentioned in Section 3. The conventional protection schemes based on the local measurements from the various buses in the microgrid system malfunction due to the data availability only from the optimal buses, that is, six buses out of 17 buses. With information acquired from the reduced number of buses, the impact of any communication-related issue is reduced. The hybrid microgrid model shown in Fig. 8.1 is simulated for AG fault in line S_2 with $R_f = 0.001\ \Omega$, $\theta_f = 90$ degrees, $L_f = 0.16$ km and line S_7 with $R_f = 0.001\ \Omega$, $\theta_f = 90$ degrees, $L_f = 0.02$ km. The current and voltage waveforms recorded at one of the optimal buses, that is, at Bus 2 corresponding to AG fault in line S_2 and S_7 are shown in Figs. 8.3 and 8.4, respectively.

The complete protection scheme is composed of two parts. The first part discusses the processing and extraction of the distinctive attributes of the voltage waveforms and current waveforms obtained at the buses 2, 6, 8, 9, 10, and 17 using DWT. In the second part, an ensemble-based classifier is executed for detecting the mode, detecting the fault, classifying the fault, and identifying the faulty line. The overall hybrid microgrid protection is explained using the flowchart as shown in Fig. 8.5.

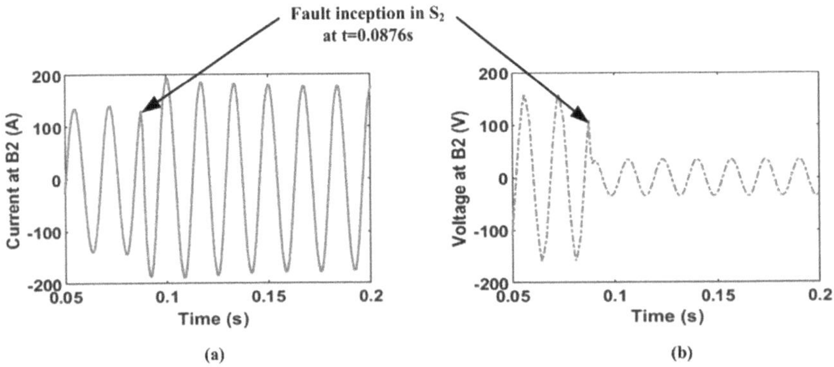

Figure 8.3 For AG fault in line S2 at 0.0876 s (A) Current waveform recorded at bus B2 (B) Voltage waveform recorded at bus B2.

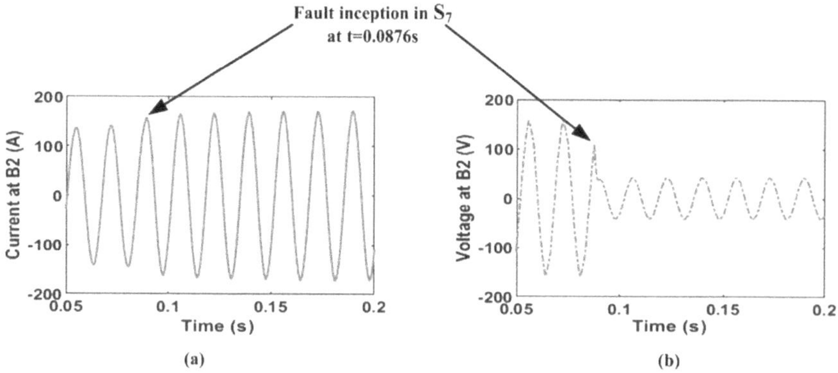

Figure 8.4 For AG fault in line S7 at 0.0876 s (A) Current waveform recorded at bus B2 (B) Voltage waveform recorded at bus B2.

4.1 Preprocessing and feature extraction using DWT

Generally, wavelet transforms are widely applied in power systems to analyze the abrupt changes in the voltage and current signals as it can provide localization in both time and frequency domain.[15] In addition to this, DWT reduces the complexity due to the presence of large data set and hence provides faster response with reduced execution time. Due to the aforementioned advantages, DWT has been implemented by the researchers in the field of power quality analysis,[25] islanding detection,[8] and designing a fault resilient microgrid protection scheme.[26]

The instantaneous values of voltage waveforms and current waveforms obtained from the six bus locations undergo preprocessing, which eliminates the redundant data (information) and extracts only the discriminative features for further analysis. The preprocessing stage involves passing the raw signals through a low-pass filter (Butterworth with cut-off frequency 480 Hz) and further sampled at 1.2 kHz. Feature extraction is done with Daubechies 3 (db3) mother wavelet as it provides the attributes with more discriminatory information for healthy and faulty scenarios.[27] The signal here is decomposed up to three levels which contains only the low-frequency components particularly the fundamental frequency component while eliminating the higher frequency components.

Decomposition of any time-domain signal into its approximate coefficients can be represented using the scaling function $\phi_{jk_2}(t)$ as[28]:

$$\phi_{jk_2}(t) = 2^{-j/2}\phi\left(2^{-j}t - k_2\right) \tag{8.3}$$

where, $k_2 \in Z$ and j is an integer value.

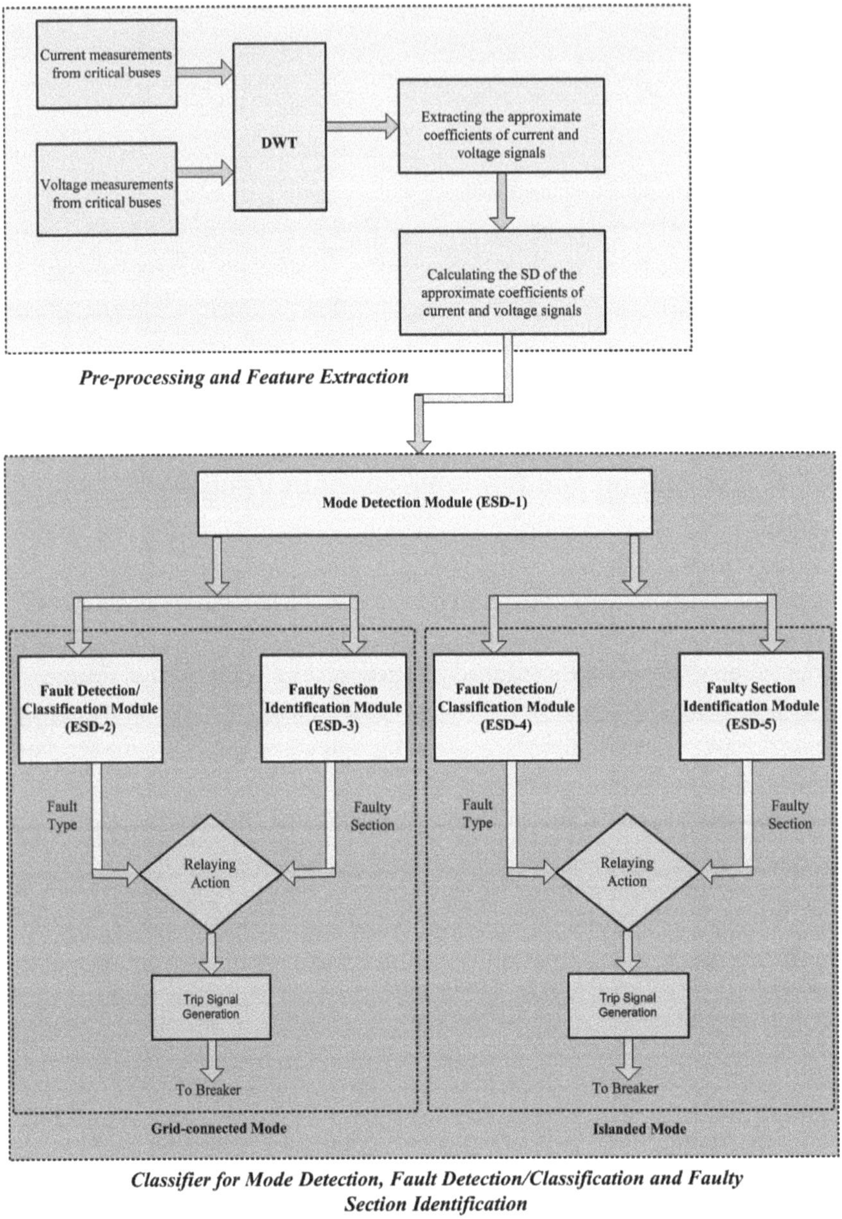

Figure 8.5 DWT and ESD-based hybrid microgrid protection scheme.

The standard deviation of third level approximate coefficients derived from current and voltage signals for the three phases (a, b, c) and for the six optimal bus locations is calculated which yields 36 distinct features. The feature vector consisting of 36 distinct attributes with $i = 1, 2, \dots \dots N$ samples is given by,

$$F_i = \{F_{i1},\ F_{i2},\ \dots \dots \dots F_{i36}\} \tag{8.4}$$

The set of feature vector is fed to the ESD classifier to perform the task of detecting the mode, detecting the fault, classifying the fault, and identifying the faulty line.

4.2 ESD-based classifier for hybrid microgrid system

Ensemble-based classifier approach has outperformed single classifier approach in the area of artificial intelligence, statistics, image processing,[29] power system protection[26], remote sensing,[30] etc., due to its ability to provide improved prediction. The methodology considered for ensemble-classifiers includes weighing and combining of individual classifiers so as to obtain a final classifier which is more robust and outperforms every single classifier. A wide variety of mechanisms for building ensemble-classifiers with different training data sets, training parameters, and also different learning algorithms have been reported in the literature.[31] Some of the popular ensemble learning algorithms include boosting, bagging, random subspace, and rotation forest algorithms.[32]

In the present chapter, the ensemble of random subspace algorithm is implemented along with linear discriminant (LD) analysis. Unlike, bagging and boosting ensemble algorithms, random subspace algorithm uses the modified feature vector. Using random subsets of components from the modified feature vector, multiple learners are constructed systematically. For improving the classification performance of the random subspace method, majority voting strategy is implemented. The classification of unlabeled data is done depending on the highest number of votes obtained by that particular class. Discriminant analysis (DA) is the classification algorithm useful for wide datasets and provides accurate and fast results as compared to other classifiers. DA uses labeled dataset to build prediction model that can be further used for classifying new set of data. DA also classifies new set of observations by identifying the boundaries which may be linear or nonlinear between groups present in the data.[33,34]LD has been used to design the protection scheme for transmission line in Ref. 35. ESD approach has not been reported in the previous work for developing a reliable, robust, and cost-effective protection scheme for hybrid microgrid systems.

The proposed protection scheme is developed using a module of ESD classifiers. The set of ESD modules are developed for detecting the mode, detecting the fault, classifying the fault, and identifying the faulty line. The fault analysis is carried out for various lines of the microgrid with varying fault parameters as shown in Table 8.1. The ESD modules corresponding to different protection tasks are discussed below.

4.2.1 Module for mode detection (ESD-1)

The current—voltage profile is extremely variable for the faults in the two microgrid operating modes (grid-connected and islanded) demands for a different protection strategy for both the modes. The mode detection module (ESD-1) has two target labels GC and IM, respectively, for the two modes.

4.2.2 Module for fault detection/classification (ESD-2 and ESD-4)

ESD-2 and ESD-4 modules detect the fault and further classify the fault in grid-connected and islanded modes, respectively. The target labels for both ESD-2 and ESD-4 are AG, BG, CG, AB, BC, AC, ABG, BCG, AG, ABC, ABCG, and 1 no-fault (n-f). The data for the n-f cases is simulated by considering load variation in the range of 30%—50%.

4.2.3 Module for faulty section identification (ESD-3 and ESD-5)

The faulty section identification module is designed with target labels S_1, S_2, S_3, ...S_{14} for both the modules ESD-3 and ESD-4, respectively.

The relaying action is taken by considering the outputs from both the modules, that is, modules ESD-2 and ESD-3 in grid-connected mode and modules ESD-4 and ESD-5 in islanding mode which corresponds to the type of fault and faulty line, respectively. Thus, a relay transfers this tripping signal to a circuit breaker in order to isolate the faulty line.

5. Performance analysis

Validation of the proposed ESD-based protection scheme is conducted for different types of distribution line faults and also by considering variation in loading conditions (no-fault cases). The testing data is generated for

Table 8.1 Fault simulation parameters for training and testing data generation.

Type of scenario	Parameters	Range	Mode of operation		Total samples
			GC	IM	
Training data					
Fault	Line faults	lg, ll, llg, lll, lllg	11	11	2664 training samples
	Fault resistance, R_f (in Ω)	0–100	2	2	
	Fault inception angle, θ_f	0–90 degrees	2	2	
	Fault location, L_f (in km)	0–6	28	28	
No-fault	Load variation	30%–50%	100	100	
Testing data					
Fault	Line faults	lg, ll, llg, lll, lllg	11	11	1420 testing samples
	Fault resistance, R_f (in Ω)	0–100	2	2	
	Fault inception angle, θ_f	0–90 degrees	2	2	
	Fault location, L_f (in km)	0–6	15	15	
No-fault	Load variation	30%–50%	50	50	

a total of 1320 fault cases and 100 n-f cases under both operating modes of microgrid as indicated in Table 8.1. A comparative analysis of the ESD-based modules has been made with the LD, support vector machine (SVM), and decision tree (DT)-based classifier modules in terms of the accuracy of all the ESD modules as shown in Fig. 8.6.

5.1 Mode detection

The mode detector ESD-1 has been tested for a total of 1420 cases. The mode detector module in the proposed protection scheme has been successful in detecting the mode of operation for all the 1420 test cases with 100% accuracy. The performance comparison has also been analyzed with LD (99.2%), SVM (99.1%), and DT (99.0%) classifiers in terms of accuracy.

5.2 Fault detection/classification

The fault detector/classifier module ESD-2 and ESD-4 has been tested for 660 fault and 50 n-f cases. Both the classifiers achieve an accuracy of 98.09% and 97.63% in grid-connected and islanded mode, respectively, as indicated in Table 8.2. In addition to accuracy, the performance comparison of the ESD based protection scheme is done with LD, SVM, and DT classifiers with regard to the dependability and security for both the modes of operation as shown in Table 8.3. The ESD classifiers are found to outperform other classifiers.

5.3 Faulty section identification

The performance of the ESD-3 and ESD-5 classifiers for the proposed scheme is compared with other classifiers such as LD, SVM, and DT as

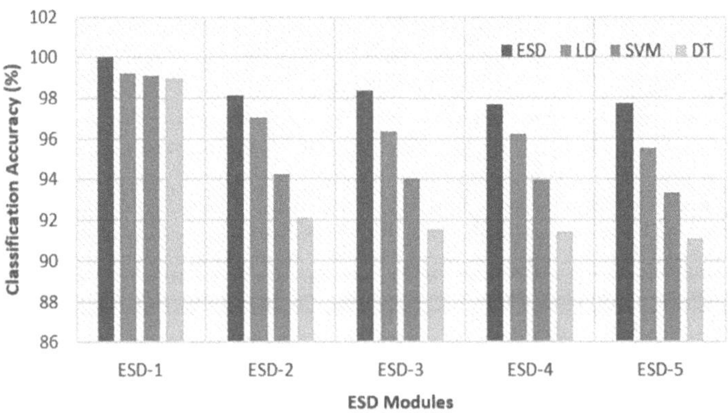

Figure 8.6 Performance comparison of proposed ESD with LD, SVM, and DT mode detection.

Table 8.2 Fault detection/classification module (ESD-2/ESD4).

Mode of operation	Fault class	lg	ll	llg	ll	llg	nf	Error (%)	Overall accuracy (%)
Output of fault detection/classification module predicted by ESD-2									
Grid-connected	lg	178	0	0	0	1	1	1.11	98.09
	ll	0	177	0	2	1	0	1.67	
	llg	1	0	177	0	1	1	1.67	
	lll	0	2	0	58	0	0	3.33	
	lllg	0	1	0	0	59	0	1.67	
	nf	0	0	0	0	1	49	2	
Output of fault detection/classification module predicted by ESD-4									
Islanded	lg	177	0	0	0	1	2	1.67	97.63
	ll	0	176	1	2	1	0	2.22	
	llg	1	0	177	0	1	1	1.67	
	lll	0	2	0	58	0	0	3.33	
	lllg	0	2	0	0	58	0	3.33	
	nf	0	0	0	0	1	49	2	

Table 8.3 Performance comparison of ESD-based scheme with other classifiers.

	Accuracy (%)		Dependability (%)		Security (%)	
Type of classifier	**GC**	**IM**	**GC**	**IM**	**GC**	**IM**
ESD	98.09	97.63	98.33	97.88	98	98
LD	97.04	96.20	97.12	96.36	96	94
SVM	94.23	93.94	94.39	94.1	92	92
DT	92.11	91.41	92.27	91.51	90	90

shown in Table 8.4. The ESD-3 and ESD-5 faulty section identification modules are capable of detecting the faulty section with an accuracy of 98.35% and 97.71% under grid-connected and islanded modes, respectively.

To assess the appropriateness of the protection scheme in providing faster relaying action against different types of faults in the microgrid system under study, two fault cases (AB and ABC) have been simulated in line S_1. For both the cases, the fault is executed at $t = 0.0876$ s , while keeping the fault resistance and location as $R_f = 0.001\ \Omega$, $L_f = 0.02$ km, same in both the cases. The current waveforms and the responses of the fault detector/classifier modules corresponding to AB and ABC fault are shown in Figs. 8.7 and 8.8, respectively. The relay generates a tripping signal for AB and ABC faults after $t = 14.9$ ms and $t = 15.73$ ms, respectively.

Table 8.4 Performance comparison of ESD-based section identifier with LD, SVM, and DT.

		Accuracy (%) of section identification							
	No. of test cases (for	ESD		LD		SVM		DT	
Sections (lines)	each mode)	GC (ESD-3)	IM (ESD-5)	GC	IM	GC	IM	GC	IM
S_1	50	98	98	96	96	94	94	92	92
S_2	40	100	100	97.5	94	94	94	92.5	92.5
S_3	50	100	100	98	96	96	94	92	90
S_4	40	97.5	95	97.5	97.5	94	92.5	92.5	92.5
S_5	50	96	96	96	96	94	92	92	92
S_6	40	97.5	97.5	94	94	92.5	92.5	90	90
S_7	40	100	97.5	97.5	94	94	94	92.5	90
S_8	50	98	96	96	96	94	92	92	90
S_9	50	100	100	98	96	94	94	92	92
S_{10}	50	96	96	94	94	94	94	90	90
S_{11}	50	98	98	96	96	94	94	90	90
S_{12}	50	100	100	96	96	94	94	92	92
S_{13}	50	98	96	96	96	92	92	90	90
S_{14}	50	98	98	96	96	96	94	92	92
Total accuracy (%)	–	**98.35**	**97.71**	**96.32**	**95.54**	**94.04**	**93.36**	**91.54**	**91.07**

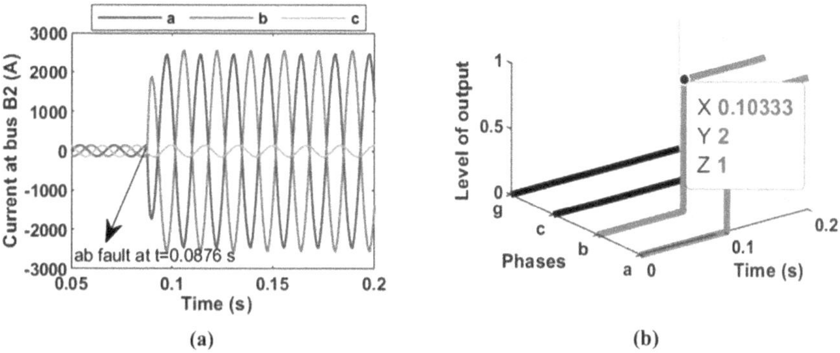

Figure 8.7 Response of the ESD-based protection scheme obtained at Bus B2 for fault in line S1 at 0.0876 s (A) Current waveforms for AB fault, (B) Response of fault detection/classification module corresponding to (A).

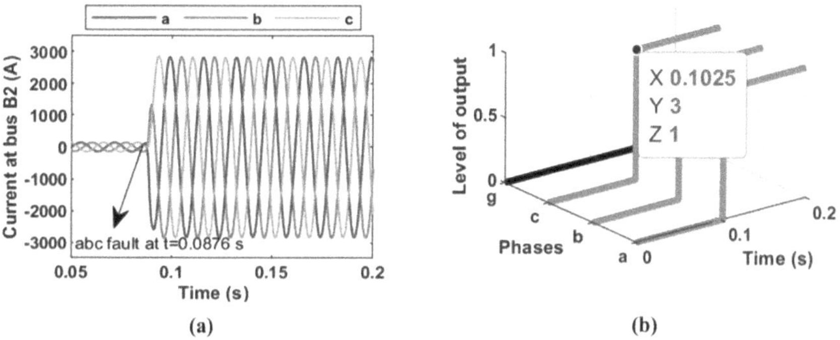

Figure 8.8 Response of the ESD-based protection scheme obtained at Bus B2 for fault in line S1 at 0.0876 s (A) Current waveforms for ABC fault, (B) Response of fault detection/classification module corresponding to (A).

6. Conclusion

Hybrid microgrids are becoming increasingly popular owing to the benefits offered by AC as well as DC microgrids. However, in hybrid microgrids, the presence of AC and DC generation along with the large number of DER units and loads makes the protection task very challenging. In addition to this, the requirement of sensors at each and every bus of the microgrid for voltage/current measurement is not feasible due to economic concerns and higher probability of sensor failure. In this context, a protection scheme using DWT and ESD is developed for a hybrid microgrid network which utilizes the data collected from the optimal number of buses in the system using

GA and observability approach. Additionally, the proposed scheme is also capable of detecting the mode, detecting the fault, classifying the fault, and identifying the faulty line for both the modes. The effectiveness of the ESD-based protection scheme is tested for wide range of operating scenarios by varying the fault parameters and at different loading conditions. The proposed ESD-based scheme has been found to outperform other machine learning-based classifiers such as LD, SVM, and DT in terms of accuracy, dependability, and security.

References

1. Gupta N, Joshal KS, Tomar A. *Environmental and Technoeconomic Aspects of Distributed Generation*. Elsevier BV; 2021:237–263. https://doi.org/10.1016/b978-0-12-824337-4.00009-6.
2. Beheshtaein S, Cuzner R, Savaghebi M, Guerrero JM. Review on microgrids protection. *IET Gener Transm Distr*. 2019;13(6):743–759. https://doi.org/10.1049/iet-gtd.2018.5212.
3. Hooshyar A, Iravani R. Microgrid protection. *Proc IEEE*. 2017;105(7):1332–1353. https://doi.org/10.1109/JPROC.2017.2669342.
4. Lotfi H, Khodaei A. AC versus DC microgrid planning. *IEEE Trans Smart Grid*. 2017; 8(1):296–304. https://doi.org/10.1109/TSG.2015.2457910.
5. Unamuno E, Barrena JA. Hybrid ac/dc microgrids—Part I: review and classification of topologies. *Renew Sust Energ Rev*. 2015;52:1251–1259. https://doi.org/10.1016/j.rser.2015.07.194.
6. Pesaran HAM, Huy PD, Ramachandaramurthy VK. A review of the optimal allocation of distributed generation: objectives, constraints, methods, and algorithms. *Renew Sustain Energy Rev*. 2017;75:293–312. https://doi.org/10.1016/j.rser.2016.10.071.
7. Gupta N. Probabilistic optimal reactive power planning with onshore and offshore wind generation, EV, and PV uncertainties. *IEEE Trans Ind Appl*. 2020;56(4):4200–4213. https://doi.org/10.1109/TIA.2020.2985319.
8. Manohar M, Koley E, Ghosh S. Reliable protection scheme for PV integrated microgrid using an ensemble classifier approach with real-time validation. *IET Sci Measur Technol*. 2018;12(2):200–208. https://doi.org/10.1049/iet-smt.2017.0270.
9. Manohar M, Koley E, Ghosh S. Microgrid protection under weather uncertainty using joint probabilistic modeling of solar irradiance and wind speed. *Comput Electr Eng*. 2020; 86:106684. https://doi.org/10.1016/j.compeleceng.2020.106684.
10. Tiwari SP, Koley E, Ghosh S. Communication-less ensemble classifier-based protection scheme for DC microgrid with adaptiveness to network reconfiguration and weather intermittency. *Sustain Energy Grids Netw*. 2021;26:100460. https://doi.org/10.1016/j.segan.2021.100460.
11. Mirsaeidi S, Dong X, Said DM. Towards hybrid AC/DC microgrids: critical analysis and classification of protection strategies. *Renew Sustain Energy Rev*. 2018;90:97–103. https://doi.org/10.1016/j.rser.2018.03.046.
12. Mirsaeidi S, Dong X, Shi S, Tzelepis D. Challenges, advances and future directions in protection of hybrid AC/DC microgrids. *IET Renew Power Gen*. 2017;11(12): 1495–1502. https://doi.org/10.1049/iet-rpg.2017.0079.

13. Baldwin TL, Mili L, Boisen MB, Adapa R. Power system observability with minimal phasor measurement placement. *IEEE Trans Power Syst*. 1993;8(2):707–715. https://doi.org/10.1109/59.260810.
14. Pal A, Sanchez-Ayala GA, Centeno VA, Thorp JS. A PMU placement scheme ensuring real-time monitoring of critical buses of the network. *IEEE Trans Power Deliv*. 2014; 29(2):510–517. https://doi.org/10.1109/TPWRD.2013.2279172.
15. Mishra DP, Samantaray SR, Joos G. A combined wavelet and data-mining based intelligent protection scheme for microgrid. *IEEE Trans Smart Grid*. 2016;7(5):2295–2304. https://doi.org/10.1109/TSG.2015.2487501.
16. Koley E, Verma K, Ghosh S. An improved fault detection classification and location scheme based on wavelet transform and artificial neural network for six phase transmission line using single end data only. *SpringerPlus*. 2015;4(1). https://doi.org/10.1186/s40064-015-1342-7.
17. Gul A, Perperoglou A, Khan Z, et al. Ensemble of a subset of kNN classifiers. *Adv Data Anal Classif*. 2018:827–840. https://doi.org/10.1007/s11634-015-0227-5.
18. Ortiz L, Orizondo R, Águila A, González JW, López GJ, Isaac I. Hybrid AC/DC microgrid test system simulation: grid-connected mode. *Heliyon*. 2019;5(12). https://doi.org/10.1016/j.heliyon.2019.e02862.
19. Gou B. Generalized integer linear programming formulation for optimal PMU placement. *IEEE Trans Power Syst*. 2008;23(3):1099–1104. https://doi.org/10.1109/TPWRS.2008.926475.
20. Clements KA. Observability methods and optimal meter placement. *Int J Electr Power Energy Syst*. 1990;12(2):88–93. https://doi.org/10.1016/0142-0615(90)90004-U.
21. Nuqui RF, Phadke AG. Phasor measurement unit placement techniques for complete and incomplete observability. *IEEE Trans Power Deliv*. 2005;20(4):2381–2388. https://doi.org/10.1109/TPWRD.2005.855457.
22. Gou B. Optimal placement of PMUs by integer linear programming. *IEEE Trans Power Syst*. 2008;23(3):1525–1526. https://doi.org/10.1109/TPWRS.2008.926723.
23. Bedekar PP, Bhide SR, Kale VS. Optimum PMU placement considering one line/one PMU outage and maximum redundancy using genetic algorithm. ECTI-CON 2011. In: *8th Electrical Engineering/Electronics, Computer, Telecommunications and Information Technology (ECTI) Association of Thailand - Conference 2011*. 2011:688–691. https://doi.org/10.1109/ECTICON.2011.5947933.
24. Müller HH, Castro CA. Genetic algorithm-based phasor measurement unit placement method considering observability and security criteria. *IET Gen Trans Distrib*. 2016; 10(1):270–280. https://doi.org/10.1049/iet-gtd.2015.1005.
25. Masoum MAS, Jamali S, Ghaffarzadeh N. Detection and classification of power quality disturbances using discrete wavelet transform and wavelet networks. *IET Sci Measur Technol*. 2010;4(4):193–205. https://doi.org/10.1049/iet-smt.2009.0006.
26. Manohar M, Koley E, Ghosh S. Stochastic weather modeling-based protection scheme for hybrid PV-wind system with immunity against solar irradiance and wind speed. *IEEE Syst J*. 2020;14(3):3430–3439. https://doi.org/10.1109/JSYST.2020.2964990.
27. Pigazo A, Liserre M, Mastromauro RA, Moreno VM, Dell'Aquila A. Wavelet-based Islanding detection in grid-connected PV systems. *IEEE Trans Ind Electr*. 2009; 56(11):4445–4455. https://doi.org/10.1109/TIE.2008.928097.
28. Vetterli M, Herley C. Wavelets and filter banks: theory and design. *IEEE Trans Signal Process*. 1992;40(9):2207–2232. https://doi.org/10.1109/78.157221.
29. Hang R, Liu Q, Song H, Sun Y. Matrix-based discriminant subspace ensemble for hyperspectral image spatial-spectral feature fusion. *IEEE Trans Geosci Remote Sens*. 2016;54(2):783–794. https://doi.org/10.1109/TGRS.2015.2465899.

30. Radhika K, Varadarajan S. Ensemble subspace discriminant classification of satellite images. *J Sci Industr Res.* 2018;77(11):633–638. http://nopr.niscair.res.in/handle/123456789/46.
31. Polikar R. Ensemble based systems in decision making. *IEEE Circuits Syst Mag.* 2006; 6(3):21–44. https://doi.org/10.1109/MCAS.2006.1688199.
32. Rokach L. Ensemble-based classifiers. *Artif Intell Rev.* 2010;33(1–2):1–39. https://doi.org/10.1007/s10462-009-9124-7.
33. Ashour AS, Guo Y, Hawas AR, Xu G. Ensemble of subspace discriminant classifiers for schistosomal liver fibrosis staging in mice microscopic images. *Health Inform Sci Syst.* 2018. https://doi.org/10.1007/s13755-018-0059-8.
34. Tavernier J, Simm J, Meerbergen K, Wegner JK, Ceulemans H, Moreau Y. Fast semi-supervised discriminant analysis for binary classification of large data sets. *Pattern Recogn.* 2019;91:86–99. https://doi.org/10.1016/j.patcog.2019.02.015.
35. Yadav A, Swetapadma A. A novel transmission line relaying scheme for fault detection and classification using wavelet transform and linear discriminant analysis. *Ain Shams Eng J.* 2015;6(1):199–209. https://doi.org/10.1016/j.asej.2014.10.005.

CHAPTER NINE

A linear discriminant analysis based protection scheme for DC microgrid under stressed scenarios

Shankarshan Prasad Tiwari, Ebha Koley and Subhojit Ghosh
Department of Electrical Engineering, National Institute of Technology, Raipur, Chhattisgarh, India

Contents

Acronyms

- DER—Distributed energy resources
- kNN—k-nearest neighbor
- LEDs—Lighting emitting diodes
- LDA—Linear discriminant analysis
- PCC—Point of common coupling

Smart Electrical and Mechanical Systems
ISBN: 978-0-323-90789-7
https://doi.org/10.1016/B978-0-323-90789-7.00010-5

- SDG—Synchronous diesel generator
- SVM—Support vector machine
- UG—Utility grid
- WTG—Wind turbine generator

1. Introduction

The demand of electrical power has seen an unpresented rise due to the modernization of the industrial infrastructure around the world. However, the classical power system bulk power generating sites are not capable of effectively solving the energy crisis because of their increased dependence on fossil fuels. The increasing energy requirement and the need of lowering the carbon footprint has led to the wider acceptance and utilization of renewable sources of energy as wind, tidal, wave, biomass geothermal, and photovoltaic generator-based PV array.[1–3] Improving operational reliability with wider integration of renewable sources has increased the focus microgrid. It is an interconnection of low rating or microscale renewable sources, loads, converters, and battery within a specified electrical boundary.[4,5] The microgrid can also be defined as a local energy active power distribution system where the voltage is generated at the distribution level. Thus, microgrids may operate by either being connected (grid-connected mode) are disconnected (islanded mode) from the utility grid. The possible connection and disconnection as per the power generation and load requirement improves the operational efficiency, reduces the impact of stressed scenarios, and avoids the possibility of blackout conditions.

The microgrid system is often categorized into three groups depending on the profile of the current/voltage at the buses: AC, DC, and hybrid microgrid systems.[6,7] In recent times, demand of DC power-based devices has risen significantly due to the its wide usage in different applications/appliances like electric traction, washing machine, air conditioners, refrigerators, savers, grinders, shipboards, residential loads LEDs, batteries, data centers. The same has led the increased adoption of DC microgrid by researchers and power engineers. The DC microgrid provides multiple advantages over the AC microgrid system like absence of synchronization problem, safer for human bodies due to less discharge, free from skin effect, eco-friendliness, higher power quality, less power losses due to less conversion stages of power, resilient, reliable, and minimal requirement of two leads only.[8]

In spite of the numerous benefits over the AC microgrid system, it is challenging to provide adequate protection to DC microgrid due to high variation in the magnitude of fault current[9] from healthy to fault scenarios. Also, the profile of postfault current is heavily dependent on the operating mode. For grid connected mode, the fault current is 10−50 times greater than islanded mode of operation. Further the disparity during faults involving only poles and pole to ground fault,[10] lack of zero crossing signals[11] and sporadic nature of renewable DERs adds to the complexity. The above issues have limited the application of the DC microgrid. Considering the importance of rapidity in detecting a fault and mitigate it's impact, the researchers have reported on a number of protective techniques for the DC microgrid system. Some of significant among them are adaptive voltage threshold-based scheme,[12] probe power unit-based noniterative approach for fault location estimation,[13] high-speed overcurrent differential protection-based scheme,[14] communication assisted unit protection scheme based on superimposed current,[15] differential current-based scheme with wavelet transform,[16] overcurrent-based scheme for ring configuration,[17] differential current,[18] communication-assisted centralized protection-based fault detection approach,[19] local measurement assisted nonunit protection,[20] communication-less localized protection scheme,[21] firefly algorithm-based protection scheme,[22] central processing unit-based fault detection,[23] and inductance estimation-based fault detection.[24] In spite of the ability of the reported schemes in offering reliable protection, majority of the techniques are not robust against the variation in the fault current's level under the random behavior of renewable sources.

Motivated by the need to address the critical issues in DC microgrid protection, in this chapter a linear discriminant analysis (LDA)-based protection algorithm is proposed to improve the system integrity and invulnerability under variation in fault parameters and stressed conditions in DC microgrid. Stressed scenarios in DC microgrid generally result from wide variation in the loading condition and random fluctuation in the weather variables (solar irradiance) leading to severe disturbance in the distribution network. The LDA algorithm because of its simple structure, high classification accuracy on complex dataset and high robustness to noisy data has found wide application in different engineering domains like face reorganization, biomedical,[25] and earth science.[26] In contrast to the more popular machine learning-based classifiers like SVM and neural networks, the tuning is generally not a tedious task, since the tuning parameters are less. For a given dataset, LDA searches for the linear combination of features/attributes that maximizes the classification accuracy.[27] The highlights of this book chapter can be summarized as:

[1] Simulation of a test DC microgrid system to show the dependence of the system behavior on the mode of operation.
[2] Development of a LDA-based machine learning algorithm to execute the protection tasks in a DC microgrid.
[3] Development of protection scheme for DC microgrid to avoid nuisance tripping during stressed microgrid operation.
[4] Evaluating the aptness of the LDA-based protection scheme under diverse system and fault variables.

The remaining part of the chapter is organized as follows: The modeling of the DC test microgrid system is described in Section 2, and LDA classifier is briefly discussed in Section 3. In Section 4, the development of the proposed protection algorithm is described while Section 5 deals with performance evaluation. Finally, summarization of the work is done in Section 6.

2. DC-microgrid architecture and fault analysis

As mentioned earlier, DC microgrid has several advantages over AC microgrid such as reduced number of power conditioning units, which reduces the losses and complication related to the control and design of power converters. Despite the benefits, the intricacies of the protection strategy have hampered the adoption of DC microgrids. The significant magnitude (as high as 10 times) and high rate of increase in fault current, grounding concerns, the lack of bipolarity, and the increased occurrence of arcs which sustain for larger period are the key challenges in DC microgrid protection. Pole to pole (P—P) and pole to ground (P—G) faults are the two types of faults in DC microgrids. The level of fault current varies depending on the fault type (PP, PG). Detection of fault and identification of faulty section are the key features of any protection scheme. Precise and fast fault detection is vital to avoid potential microgrid component damage because of the high magnitude fault current during fault. Thus, DC microgrid protection necessitates a faster of response time because the high magnitude fault current has a damaging impact on the converter switches. In this regard, the development of a reliable and fast protection scheme is sought in this chapter, and Figs. 9.1 and 9.2 show the variance in line current information during PP and PG faults in islanded and grid linked mode.

3. DC test microgrid system under study

Fig. 9.3 illustrates the DC microgrid considered in the present work.[17] The proposed system consists of the six buses from A1 to A6 as well as six sections from S1 to S6. The system has a total length of 6 km, with each

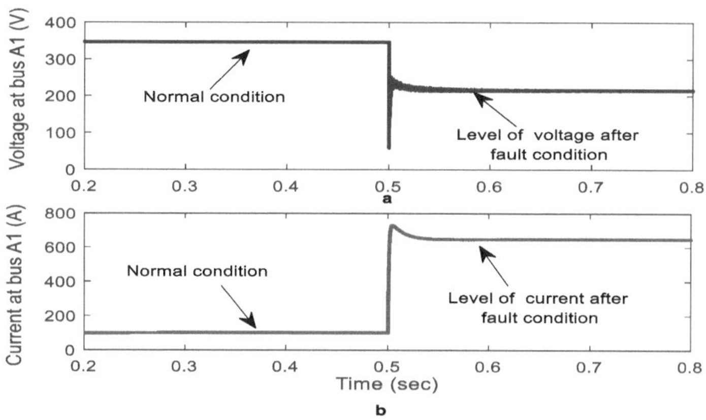

Figure 9.1 Variation in current dynamic under islanded mode during (A) PG and (B) PP fault Current variation during fault in islanded mode.

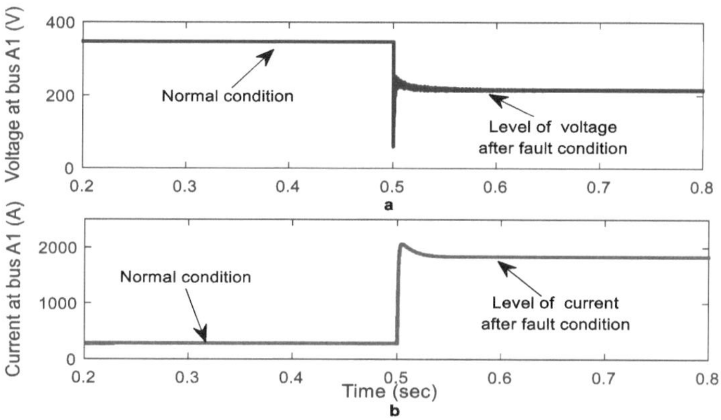

Figure 9.2 Variation in current dynamic under grid-connected mode during (A) PG and (B) PP fault Variation in current dynamic under grid-connected mode.

section spanning 1 km in length. In this test microgrid system, three DERs, namely, PV array, SDG, and WTG are connected at bus A3, A4, and A6, respectively. Apart from these sources, an energy storage device is also connected at bus A5 to improve the stability of the system. Similar to the DERs, the utility grid is also integrated at bus A6 with help of a converter and PCC (point of common coupling). A total number of six faults from F1 to F6 have also been simulated in the respective sections. Two loads, namely, L1 and L2 are also connected at the bus A2 and A4, respectively. The MATLAB/Simulink software is utilized to simulate the scenarios of the faulty conditions under both modes of operation.

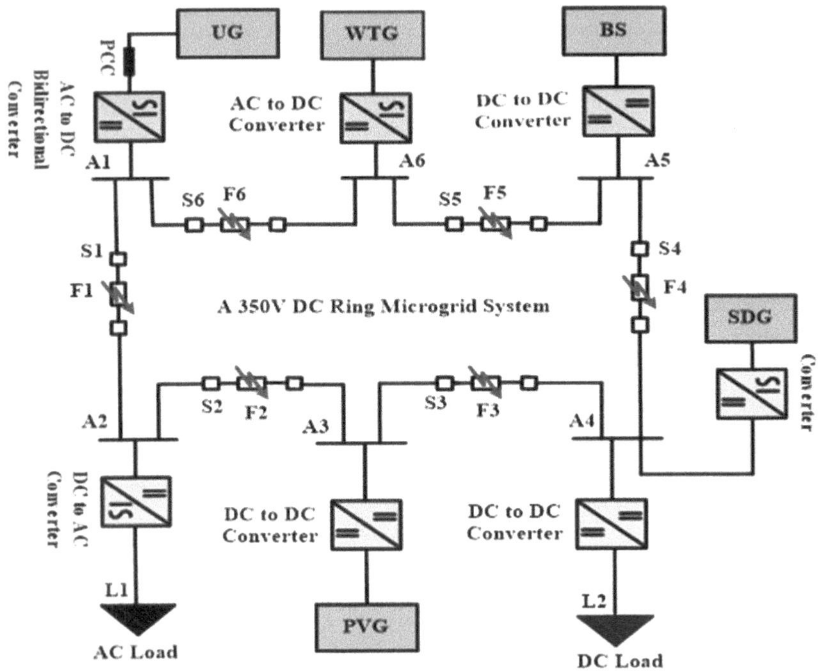

Figure 9.3 DC microgrid system Microgrid model used in the study.

4. Linear discriminant analysis-based protection scheme

The LDA is an emerging machine learning classifier. In recent years, it has been extensively used solving complex classification problems. For multidimensional datasets, it is helpful in the dimensionality reduction in some defined targets. During the classification of the data by the classifier, a fitting function is utilized to train the classifiers and it estimates the parameters using the Gaussian distribution function for each class. The classifier predicts the class with the lowest misclassification cost to forecast the class for new data.

In addition to classification, it is also been used for regression-based problems. Fig. 9.4 displays the separation of data points by an LDA classifier in two-dimensional plane. While searching for the linear separability, LDA projects the features from a high dimensional to low dimensional space. During the projection, the redundant components in the feature space are avoided to improve the classification accuracy.

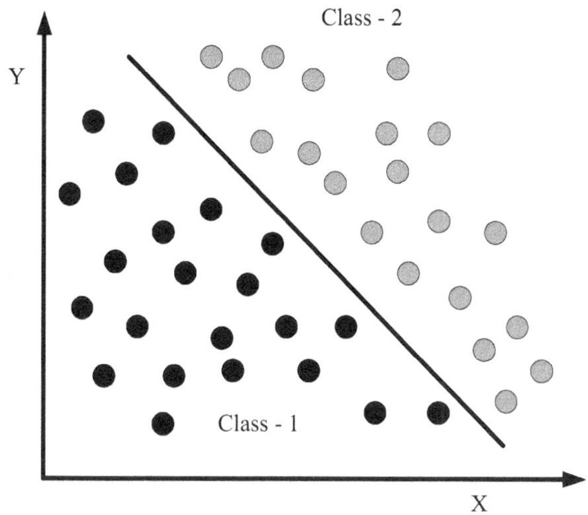

Figure 9.4 The basic principle of LDA-based algorithm Two class classification using LDA.

5. Flowchart of the proposed LDA-based scheme

The execution of the different stages of the algorithm after acquiring the sampled signals is represented in the form of a flowchart in Fig. 9.5. The overall scheme comprises of five classifier modules. Each module is dedicated to execute different tasks pertaining to overall protection of the DC microgrid. The sequence of the operations is initiated with the retrieval of postfault samples from bus A1 and further feeding it to LDA-1 for determining the operating mode. It is attained by retrieving the samples from bus A1 are feeding to the mode detector LDA-1. Once the mode is successfully detected, the algorithm triggers the subsequent LDAs. For grid linked mode, the classifier modules LDA-2 and LDA-4, detect the fault, and identify the infected section, while for islanded mode the same is performed by LDA-3 and LDA-5, respectively. The LDAs developed for fault detection and classification tasks provide output PP, PG or NF representing P—P fault, P—G fault and no fault, respectively. Similarly, LDAs developed for section identification provides output corresponding to the section in which fault has occurred. A signal is communicated to trip the circuit breaker for necessary troubleshooting if the fault is detected and the faulty section is recognized in each mode.

To examine the proposed protection scheme, an extensive set of simulated test cases have been executed by the LDA classifier modules, and the

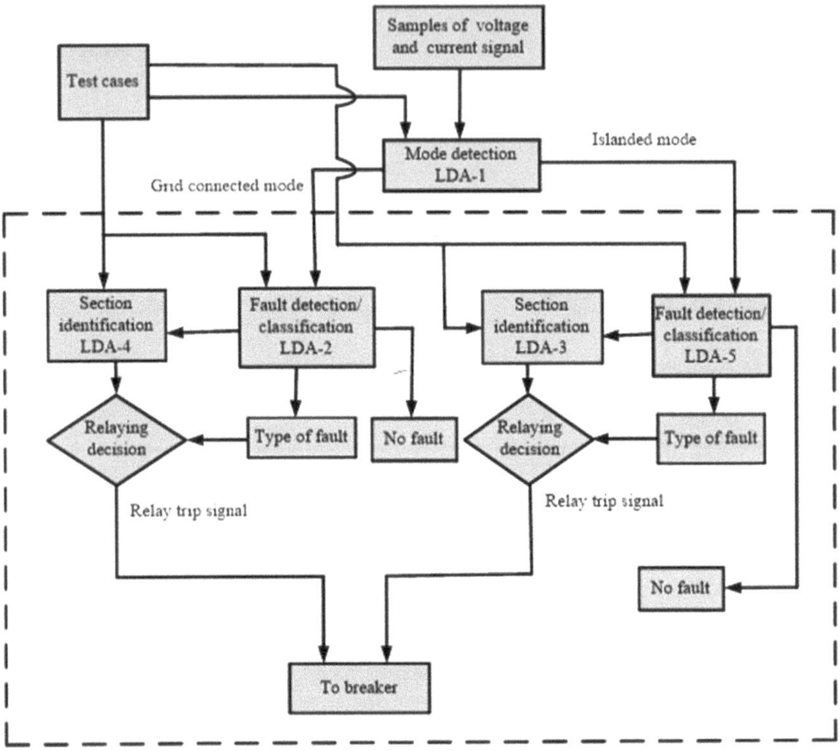

Figure 9.5 LDA-based protection scheme Proposed LDA-based DC microgrid protection scheme.

performance quantifies by the standard reliability indices. To provide the protection scheme sturdiness against varying operating and fault scenarios, care has been taken to include the test samples under varying fault parameters and weather condition in the dataset generated for designing the classifier. Further, the immunity of the any type of protection to stressed scenario can be enhanced only if it can effectively perform under wide dissimilarities in the operating condition of the distribution system. Therefore, to gauge the effectiveness of the LDA-based algorithm, many diverse test cases under the stressed conditions have also been considered. The simulated time domain signals are fed to the all respective LDAs, and the percentage accuracy is calculated after the modules have been executed. The test results are reported and analyzed in the next section.

6. Performance validation

The proposed scheme is validated to ensure its appropriateness for different operational condition of the microgrid. High accuracy of any protection scheme positively impacts the stability and efficacy of the system. For testing the classification accuracy of the LDA classifier modules, 1834 cases have generated for training and testing of the protection scheme under diverse conditions. These fault parameters varied for generating the datasets include are location of fault, fault resistance, and inception time of fault. From the above generated cases, the datasets are segregated into training and validation sets. Table 9.1 shows the generated data for the intended protective functions in both modes of the DC microgrid system.

Further, to demonstrate the change in the current and voltage dynamics after the inception of fault, the microgrid model is simulated, where the fault is created at 0.5 s and run the model for time duration of the t = 0.8 s. The simulation results are depicted in Fig. 9.6. It can be observed that following the inception of the fault, the magnitude of the voltage is falling from the normal level while the fault current rises rapidly. It is also observed that the deviation in the fault level between fault and healthy condition is more in grid-connected mode. Fig. 9.7depicts the trip signal generated for the fault cases (Fig. 9.6).

6.1 Mode detection

In this subsection, the performance of the proposed mode detector is described in terms of the classification performance of the classifier module LDA-1. To investigate the classification accuracy, 385 test cases are considered. The testing results are depicted in Table 9.2. The proposed mode detector based on LDA performs superior than other widely used classifiers such as k-nearest neighbor (kNN) and decision tree (DT) for the same datasets. To illustrate the disparity in the percentage accuracy of different classifiers a bar graph is plotted in Fig. 9.8.

6.2 Performance evaluation of LDA classifier modules for fault detection/classification

Rapidity in achieving the pre-fault scenario at the earliest, demands prompt and accurate detection and classification of faults. Accurate fault detection and classification, significantly contributes to the overall reliability of the microgrid system. In this regard, the performance of the LDA modules LDA-2 and LDA-5 have been quantified in terms of security and dependability. The indices ensure the robustness of the scheme on the basis of the higher accuracy. Dependability and security respectively quantify the ability of the protection scheme to selectively trip for fault cases and avoid unintended trips for healthy cases.

Table 9.1 Training and testing pattern for proposed algorithm.

Training parameters	Varying case of fault parameters	Total number of generated fault cases	Number of training cases	Number of testing cases
Types of mode	2	1584 fault+250 no fault cases = 1834	1284	550
Section	6			
Types of fault	2			
Fault resistance (Ω)	11			
Length of fault (km)	11			
Inception time of fault	1			

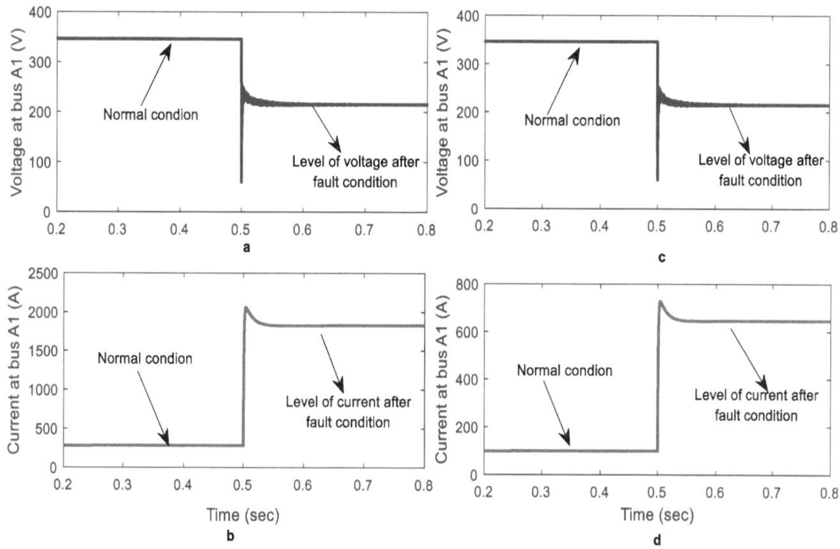

Figure 9.6 Variation in the current and voltage profile during fault conditions (A) and (B) for grid connected mode, (C) and (D) for islanded mode Variation in the current and voltage profile during fault conditions for grid connected mode, and islanded mode.

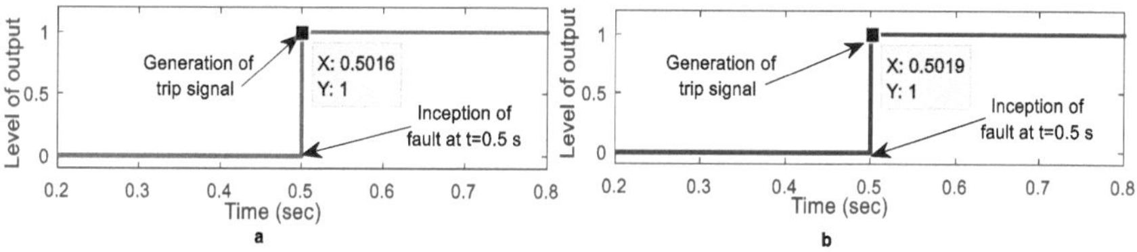

Figure 9.7 Trip signal generated for faults in Fig. 11.6 (A) and (B) for grid connected mode (C) and (D) for islanded mode Trip signal generated for faults in Fig. 9.6 (A) for grid connected mode, (D) for islanded mode.

Two widely used machine learning methods kNN (k-nearest neighbor) and DT (network) are compared with the proposed scheme. The greater accuracy reveals that the proposed fault detector is more accurate and efficient in detecting and classifying faults in both grid and islanded mode (Table 9.3).

6.3 Identification of section involved in fault

The identification of the section infected in the fault is also an important task, since it allows for isolating the faulty section. In this subsection, the

Table 9.2 Comparison of the performance of the mode detector with standalone kNN and DT.

Technique	Type of mode		
	Accuracy (%) grid connected	**Accuracy (%) islanded**	**Overall accuracy (%)**
LDA	98.96	97.88	98.42
Standalone kNN	96.78	95.76	96.27
DT	98.89	97.92	98.4

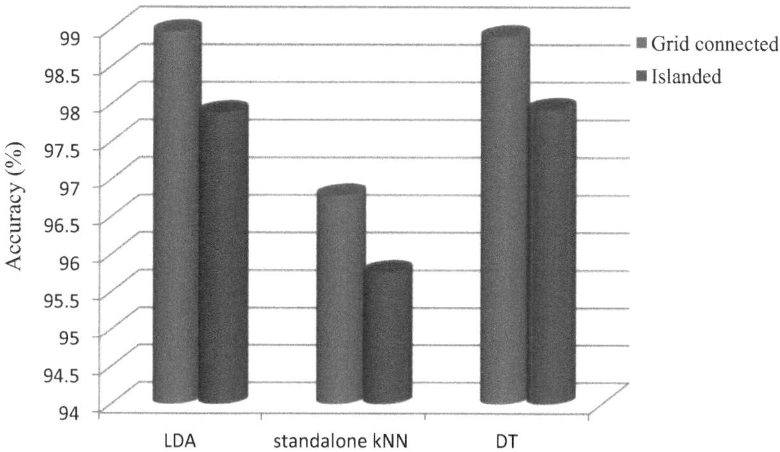

Figure 9.8 Comparison of the accuracy with the proposed algorithm and other machine learning based classification algorithms.

accuracy of the section identifier has been analyzed. Table 9.4 depicts the response for pole to ground faults and pole to pole faults, respectively, under grid connected and islanded mode. All the sections from D1 to D6 are considered for evaluating the effectiveness of the proposed section identifier. The percentage accuracy indicates the overall efficacy of the section identifier. The desired task of section identification is performed using 360 events, and the same datasets are used for the two other section identifiers as well.

6.4 Classifier modules response under the variation in the fault resistance

The response of the proposed algorithm is examined under the variation of the fault resistances for both PP and PG faults. The response of the proposed protection scheme is depicted in Table 9.5 where the value of the fault resistances is widely varied. The test results indicate that the output of the classifier modules is not impacted by the change in fault resistance.

6.5 Response of the proposed scheme under the variation in the stressed scenarios

The response of the proposed algorithm is examined under stressed microgrid condition such as switching on/off load and change in the weather-dependent dynamics of DER (PV). The response of the LDA-based classifier modules is depicted in Tables 9.6 and 9.7 where the load profile is widely

Table 9.3 Performance evaluation of the LDA classifier modules LDA-2 and LDA-5.

	Type of mode					
	Grid-connected			Islanded		
Technique	**Dependability(%)**	**Security(%)**	**Accuracy(%)**	**Dependability(%)**	**Security(%)**	**Accuracy(%)**
Proposed algorithm	99.11	100	99.55	98.36	100	99.18
Standalone kNN	97.76	96.16	96.96	98.22	97.16	97.69
DT	97.22	98.23	97.72	97.42	96.56	96.99

Table 9.4 Section identification accuracy of LDA-based algorithm for PP and PG faults.

Operational mode	Section	Proposed algorithm (accuracy %)	Standalone kNN (accuracy %)	DT (accuracy %)
Grid-connected mode	S1	98.66	95.33	96.36
	S2	98.32	96.23	97.22
	S3	99.61	95.68	96.49
	S4	98.65	94.43	96.85
	S5	98.76	95.46	97.23
	S6	99.22	97.64	97.81
	Overall accuracy	98.87	95.79	96.99
Islanded mode	S1	97.94	94.36	94.62
	S2	97.44	95.22	95.52
	S3	98.33	94.36	96.12
	S4	98.14	94.12	95.26
	S5	97.26	95.16	95.98
	S6	98.12	94.36	96.86
	Overall accuracy	97.87	94.6	95.72

Table 9.5 Response of the LDA-based algorithm against the varying load conditions.

Type of mode	Type of fault	Fault resistance (Ohm)	Faulty section	Response of LDA classifier
Grid-connected	PG	25	S1	PG
	PG	35	S6	PG
	PG	45	S3	PG
	PG	55	S4	PG
	PP	0.4	S3	PP
	PP	0.4	S2	PP
	PP	0.4	S5	PP
	PP	0.4	S6	PP
Islanded	PG	15	S1	PG
	PG	30	S6	PG
	PG	45	S3	PG
	PG	60	S4	PG
	PP	0.4	S3	PP
	PP	0.4	S2	PP
	PP	0.4	S5	PP
	PP	0.4	S6	PP

Table 9.6 Response of the LDA-based algorithm under the load variation and the solar intermittency (Islanded mode).

				Response of LDA	
Fault type	**Load variation (%)**	**Faulty section**	**Irradiance (W/m²)**	**Fault type**	**Faulty section**
PG	+20	S2	200	PG	S2
	−20	S4	300	PG	S4
	+40	S6	400	PG	S6
	−40	S4	500	PG	S4
PP	+20	S1	500	PP	S1
	−20	S3	300	PP	S3
	+40	S6	600	PP	S6
	−40	S5	200	PP	S5
Non faulty	+30, −30	—	300	Non faulty	—
	+20, −20	—	500	Non faulty	—

Table 9.7 Response of the LDA-based algorithm under load variation and solar intermittency (Grid connected mode).

Fault type	Load Variation (%)	Faulty Section	Irradiance variation (W/m^2)	Response of LDA: Fault type	Response of LDA: Faulty Section
PG	+20	S2	200	PG	S2
	-20	S4	300	PG	S4
	+40	S6	400	PG	S6
	-40	S4	500	PG	S4
PP	+20	S1	500	PP	S1
	-20	S3	300	PP	S3
	+40	S6	600	PP	S6
	-40	S5	200	PP	S5
NF	±30	—	—	NF	—
	±20	—	—	NF	—

varied. Along with load variation, the PV irradiance has also been varied. The validation results indicate that the overall performance of proposed scheme is not affected with variation in stressed scenarios such as load variation and intermittency related to PV irradiance level.

7. Conclusions

In this chapter, a LDA-based algorithm has been proposed to protect DC microgrid. The proposed LDA-based scheme enhances the system integrity and immunity under variation in fault parameters and stressed conditions in DC microgrid. The raw signals measured at the PCC are utilized to train and test the proposed algorithm. The technique has been thoroughly tested in a variety of fault circumstances. In addition to accurately detecting the fault, the scheme is able to enhance the reliability of the overall system, by ensuring a high level of dependability and security.

References

1. Adefarati T, Bansal RC. Integration of renewable distributed generators into the distribution system: a review. *IET Renew Power Gener.* 2016;10(7):873–884. https://doi.org/10.1049/iet-rpg.2015.0378.
2. Gupta N. Probabilistic optimal reactive power planning with onshore and offshore wind generation, EV, and PV uncertainties. *IEEE Trans Ind Appl.* 2020;56(4):4200–4213. https://doi.org/10.1109/TIA.2020.2985319.
3. Gupta N, Joshal KS, Tomar A. *Environmental and Technoeconomic Aspects of Distributed Generation.* Elsevier BV; 2021:237–263. https://doi.org/10.1016/b978-0-12-824337-4.00009-6.
4. Hooshyar A, Iravani R. A new directional element for microgrid protection. *IEEE Trans Smart Grid.* 2018;9(6):6862–6876. https://doi.org/10.1109/TSG.2017.2727400.
5. Manohar M, Koley E, Ghosh S. Microgrid protection under wind speed intermittency using extreme learning machine. *Comput Electr Eng.* 2018;72:369–382. https://doi.org/10.1016/j.compeleceng.2018.10.005.
6. Kamran M, Bilal M, Mudassar M. DC home appliances for DC distribution system. *Mehran Univ Res J Eng Technol.* 2017:881–890. https://doi.org/10.22581/muet1982.1704.12.
7. Mirsaeidi S, Dong X, Said DM. Towards hybrid AC/DC microgrids: critical analysis and classification of protection strategies. *Renew Sustain Energy Rev.* 2018;90:97–103. https://doi.org/10.1016/j.rser.2018.03.046.
8. Cairoli P, Rodrigues R, Zheng H. Fault current limiting power converters for protection of DC microgrids. *Conf Proc IEEE Southeastcon.* 2017. https://doi.org/10.1109/SECON.2017.7925392.
9. Yang J, Fletcher JE, O'Reilly J. Short-circuit and ground fault analyses and location in VSC-based DC network cables. *IEEE Trans Ind Electron.* 2012;59(10):3827–3837. https://doi.org/10.1109/TIE.2011.2162712.
10. Dhar S, Dash PK. Differential current-based fault protection with adaptive threshold for multiple PV-based DC microgrid. *IET Renew Power Gener.* 2017;11(6):778–790. https://doi.org/10.1049/iet-rpg.2016.0577.

11. Sarangi S, Sahu BK, Rout PK. Distributed generation hybrid AC/DC microgrid protection: a critical review on issues, strategies, and future directions. *Int J Energy Res*. 2020;44(5):3347−3364. https://doi.org/10.1002/er.5128.
12. Mohammadi J, Badrkhani Ajaei F. Adaptive voltage-based load shedding scheme for the DC microgrid. *IEEE Access*. 2019;7:106002−106010. https://doi.org/10.1109/ACCESS.2019.2932327.
13. Mohanty R, Balaji USM, Pradhan AK. An accurate noniterative fault-location technique for low-voltage DC microgrid. *IEEE Trans Power Deliv*. 2016;31(2):475−481. https://doi.org/10.1109/TPWRD.2015.2456934.
14. Fletcher SDA, Norman PJ, Fong K, Galloway SJ, Burt GM. High-speed differential protection for smart DC distribution systems. *IEEE Trans Smart Grid*. 2014;5(5):2610−2617. https://doi.org/10.1109/TSG.2014.2306064.
15. Mohanty R, Pradhan AK. A superimposed current based unit protection scheme for DC microgrid. *IEEE Trans Smart Grid*. 2018;9(4):3917−3919. https://doi.org/10.1109/TSG.2018.2835645.
16. Som S, Samantaray SR. Efficient protection scheme for low-voltage DC micro-grid. *IET Gener Transm Distrib*. 2018;12(13):3322−3329. https://doi.org/10.1049/iet-gtd.2017.1533.
17. Mohanty R, Pradhan AK. Protection of smart DC microgrid with ring configuration using parameter estimation approach. *IEEE Trans Smart Grid*. 2018:6328−6337. https://doi.org/10.1109/tsg.2017.2708743.
18. Park JD, Candelaria J, Ma L, Dunn K. DC ring-bus microgrid fault protection and identification of fault location. *IEEE Trans Power Deliv*. 2013;28(4):2574−2584. https://doi.org/10.1109/TPWRD.2013.2267750.
19. Monadi M, Gavriluta C, Luna A, Candela JI, Rodriguez P. Centralized protection strategy for medium voltage DC microgrids. *IEEE Trans Power Deliv*. 2017;32(1):430−440. https://doi.org/10.1109/TPWRD.2016.2600278.
20. Meghwani A, Srivastava SC, Chakrabarti S. A non-unit protection scheme for DC microgrid based on local measurements. *IEEE Trans Power Deliv*. 2017;32(1):172−181. https://doi.org/10.1109/TPWRD.2016.2555844.
21. Bayati N, Hajizadeh A, Soltani M. Localized fault protection in the DC microgrids with ring configuration. *IEEE Int Symp Ind Electron*. 2019;2019:136−140. https://doi.org/10.1109/ISIE.2019.8781176.
22. Amamra SA, Ahmed H, El-Sehiemy RA. Firefly algorithm optimized robust protection scheme for DC microgrid. *Elec Power Compon Syst*. 2017;45(10):1141−1151. https://doi.org/10.1080/15325008.2017.1319435.
23. Madingou G, Mahyar Z, Vaziri M. Fault detection and isolation in a DC microgrid using a central processing unit. *IEEE Power Energy Soc Innova Smart Grid Technol Conferen*. 2015:1−5.
24. Shamsoddini VB, Razani R, Yasser A. A novel protection scheme for low voltage DC microgrid using inductance estimation. *Int J Electr Power Energy Syst*. 2020:120.
25. Slimen IB, Boubchir L, Mbarki Z, Seddik H. EEG epileptic seizure detection and classification based on dual-tree complex wavelet transform and machine learning algorithms. *J BioMed Res*. 2020;34(3):151−161. https://doi.org/10.7555/JBR.34.20190026.
26. Bhardwaj A, Singh MK, Joshi PK, et al. A lake detection algorithm (LDA) using landsat 8 data: a comparative approach in glacial environment. *Int J Appl Earth Obs Geoinf*. 2015:150−163. https://doi.org/10.1016/j.jag.2015.01.004.
27. Ciabattoni L, Cimini G, Ferracuti F, Freddi A, Ippoliti G, Monteriù A. A novel LDA-based approach for motor bearing fault detection. In: *Proceeding − 2015 IEEE International Conference on Industrial Informatics, INDIN 2015*. 2015:771−776. https://doi.org/10.1109/INDIN.2015.7281834.

CHAPTER TEN

Machine learning application to power system forecasting

B. Rajanarayan Prusty[5], Kishore Bingi[5], G. Arunkumar[1], C. Dhanamjayulu[1], Neeraj Gupta[6], Anuradha Tomar[3], Rakesh Sehgal[2]

[1]Vellore Institute of Technology, Vellore, India
[2]National Institute of Technology, Srinagar, India
[3]Netaji Subhas University of Technology, Delhi, India
[4]Assistant Professor, School of Electrical Engineering, Vellore, Tamil Nadu, India
[5]School of Electrical Engineering, Vellore Institute of Technology, Vellore, Tamil Nadu, India
[6]Department of Electrical Engineering, National Institute of Technology, Srinagar, Jammu and Kashmir, India

Contents

1. Introduction

Out-of-sample prediction, referred to as forecasting, is helpful in power system planning and operational activities as increasing uncertainties are arising due to higher penetration of renewable generations and other input random variables (RVs).[1,2] Probabilistic steady-state forecasting is increasingly crucial for quantifying uncertainties in power system decision variables.[3] Since the system planner and operator get probabilistic information regarding system steady-state beforehand, an essential advantage of assisting

Smart Electrical and Mechanical Systems
ISBN: 978-0-323-90789-7
https://doi.org/10.1016/B978-0-323-90789-7.00013-0

effective power balance at any time instants in a day can be accomplished, thereby better managing network congestions.[3] Unlike point forecasting, probabilistic forecasting can provide probability distributions in terms of probability density functions (PDFs) or cumulative distribution functions (CDFs) of input RVs for a given forecasting horizon.[4–7] Further, accurate forecasting reduces operating costs, improves system reliability, and allows efficient load management. In power system analyses, the frequently used RVs are load power, renewable generations, ambient temperature, etc.[2,8–10] Compared to the probabilistic forecasting of the load and ambient temperature, renewable generation forecasting is much more challenging as weather conditions are stochastic. The greater the uncertainty between the actual values and the predicted values, the less confident the power engineer will rely on forecasts for maintaining system reliability, especially in higher penetration of renewable generation scenarios. The use of machine learning (ML) algorithms for forecasting has gained rapid interest in the recent decade.[11,12] They outperform approaches based on Box–Jenkins methodology, as evident from various research studies. Recently increasing interest is put on methods based on deep learning (DL) and developing new architectures. Unfortunately, most of the ML and DL approaches yield point (or mean) prediction results. The intricate mathematical treatment is essential for yielding probabilistic forecast results using ML and DL methods.

The popular probability models for power system input RVs are parametric probability distribution functions, for example, Gaussian, Beta, Weibull, etc.[8,13] In reality, input RVs have varying probability distribution patterns at different time instants, places, and forecast time scales, and therefore a specific parametric distribution assumption is not realistic.[9,10] Thanks to nonparametric approaches, those estimate the predictive densities directly from the data without any constraints on the distribution shape.[5] This uncertainty narrows down while moving from planning to operational time scale.

2. Forecasting for probabilistic power system planning

Adopting probability concepts in power system forecasting allows reflecting underlying uncertainties better than deterministic approaches—further assists in accomplishing efficient power system planning and operation. Probabilistic forecasts quantify the uncertainty associated with the point forecast. With additional historical data, the forecasting performance can be improved. Historical data are used to generate a location-specific, time-of-year dependent probability distribution for input RVs. A probabilistic forecast is defined as the forecasting at hour "h," the PDF (or equivalently, the CDF) at "h + k" using a definite number of samples measured until hth hour as the input data, where $k = 1, 2, \ldots, N_{hour}$. For an hour ahead analysis, $k = 1$. Modeling of planning uncertainty considers all possible past values of inputs. Therefore, quantification of such uncertainties is decisive regarding investments and their economic merits.[3] Literature on probabilistic uncertainty characterization of power systems inputs suitable for planning application is detailed underneath.

The daily electricity demand time-series majorly influenced by variation in weather variables are wreaked due to climate change.[14] A yearly demand pattern that is significantly different from the coldest day in winter to the hottest day in summer characterizes the annual variation entirely is periodic. Such a volatile electricity demand time-series is very much interested in planning a new power plant and transmission augmentation. Accurate long-term daily peak electricity demand forecasting is essential for the national electricity supply system.[15] The over forecasted value causes several generators to start leading to surplus power reserve, whereas underforecasting fails to provide the required power during the peak demand periods. Additionally, peak electricity demand forecasting also acts as a fundamental step for power system generation planning.[16] Time factors, such as time of day, day of the week, time of year, and public holidays, influence electricity consumption. They are the major deciding factors for the development of the forecasting model. For long-term planning, deciding on the time of the year conforming to the peak electricity demand time-series apt for the context is imperative. A rich body of literature exists on point forecasting of peak electricity demand. For expansion planning,[17] the expected value and the corresponding confidence interval of the future values are forecasted. The daily peak loads of substations (each day at the same time) for a period of few months (typically, one to 3 months) corresponding to the annual peak period are considered for uncertainty modeling. Unlike the conventional model,[17] only peak load powers are considered the worst-case condition. The load power uncertainty model in Ref. 18, on the other hand, can better assess a power system's possible state. While peak load values cause low voltage and generation deficiency, light loads give rise to overvoltage conditions. A multitime instant uncertainty model for aggregate load power is discussed in Ref. 10.

Solar PV generation is cost-effective and easy to commission among various renewable generation sources. The PV generation output changes uncontrollably as the Sun rises and sets in and clouds pass over PV systems.[6]

It is site-specific and has daily and seasonal patterns. The production depends on global irradiance incident on the panel, which is not uniform over time. In the short-term, solar energy is rich in the day time than in night time, whereas in the long-term, solar energy is rich in spring than in winter. It has following major characterizing features: (1) *predictability* (expected change), (2) *uncertainty/randomness* (unexpected change), (3) *intermittency* (unplanned unavailability), (4) *uncontrollability* (power output is not defined by system management), and (5) *nondispatchable* due to (2) and (3). The uncertainty refers to the change in radiation that is mainly derived from clouds that block the Sun's radiation and grant PV power forecasting a certain level of uncertainty. The uncertainty component is either due to a quick-acting phenomenon such as cloud, fog, etc. (results in the drop of power) or due to cloud-edge effect (causes increase in power). Here precise forecasting is of the greatest importance for higher PV penetrations. These characterizing features pose significant planning and operational challenges[3]; hence, probabilistic PV generation forecasting is essential, especially for higher penetration levels. The variance is a metric for PV generation variation. PV generation is variable at different time scales; hourly, daily, and seasonally. Predictability refers to the extent to which the output power can be inferred in advance (at all the time scales, e.g., for the next hour, day, week, month, or year). For a specific time of the day, the pattern PV generation variation follows across the year, representing its variability.[19] Variability, in general, predictable variation of PV generation is the expected change in generation value. Understanding PV variation helps planners and operators develop effective measures to manage variability at different levels of PV penetration. PV generation forecasting has always been researching interest in the study of PV integrated power systems. Further, there is a rich body of literature on quantile regression (QR)-based probabilistic PV generation forecasting.[5,20,21] A probabilistic PV forecast is to estimate the probabilistic distribution of PV generation at a future time. Therefore, from a long-term perspective, accurate probabilistic forecasting of PV generation would be meaningful for PV integrated power system planning. The importance of probabilistic forecasting for other power system inputs such as wind power generation and ambient temperature, etc. is well recognized in the literature, and various probabilistic models were suggested for planning applications. Detailed modeling steps can be found in Refs. 9,10,22.

3. Data preprocessing steps

A prediction model development necessitates historical records of the target variable. The recorded dataset's correctness decides the accuracy of the forecasted output.[23] The presence of missing samples and outliers in raw data makes it unsuitable for its direct use for forecasting model development. Therefore, data preprocessing (or data cleaning or data cleansing) is used to obtain a suitable dataset to develop a prediction model. Among the various, two essential and typical steps in data preprocessing are (1) resolving inconsistencies by imputing missing values and (2) detection of outliers and their apt correction.[23–26] Therefore, the two significant steps include data preprocessing and a suitable prediction model for forecasting.

Resolving inconsistencies in raw data is an important step to retain the time series property of the dataset. A straightforward strategy frequently adopted in the literature is to replace the missing samples with the recent previous values. On the other hand, samples with unexpected spikes or dips in a time series that are unusual and inconsistent with the rest of the dataset are referred to as outliers.[23] They are alternatively known as outlying observations, discordant observations, aberrant samples, anomalies, abnormal data, deviants, or contaminants. They are randomly present in the dataset. The deviation with which one outlier is situated from its expected position value may differ from that of the other outliers. Estimation of parameters of a prediction model is often biased due to the presence of outliers and therefore yields biased forecast output. Thus, outlying observations in time series data should be detected and corrected to obtain improved forecasting results. The main challenges faced include

- Performing faster preprocessing
- Performing accurate preprocessing in all parts of time series
- Deal with the large volume of data

Outlier detection is typical of interest in a time series' local subsections. It includes determining the location, type, and magnitude of outliers present in a time series. Besides, outlier correction or outlier replacement is an equally important task like detection. Numerous methods have been proposed in the literature for outlier detection and correction.[23–26] While distance-based and statistical methods are applied to univariate time series, cluster- and density-based methods, on the other hand, are suitable for multivariate datasets.[23] As compared to the latter set of methods, statistical methods such as B-spline smoothing-based method,[24] portrait dataset-based (PDB) method,[26] improved sliding window prediction (ISWP) method,[25] etc. have gained increasing interest in power system applications.

Several studies have examined the above statistical methods' capability to handle intricate trend effect, seasonality, and volatility in time series data. Further, the way one method handles the above effects differ from others was also studied. The performance comparison is made in the literature considering polluted data (formed by introducing outliers randomly in the clean dataset) using suitable metrics.[26] Based on these studies, PDB and ISWP are two befitting methods to preprocess a time series with the effects above. They compute confidence interval/outlier region for outlier detection. PDB is a data visualization technique principally based on periodic patterns. Once the given time series is converted into portrait datasets, each portrait dataset's outlier region is calculated based on its suitable parametric distribution fit. On the other hand, instead of any user-defined threshold, predicted confidence interval (PCI) is used in ISWP for outlier detection. PCI is calculated dynamically according to different nearest-neighbor-windows sizes and confidence coefficients supplied by the user. They are using only a few recent values (which have high correlation) to predict the value at a time instant and then use that predicted value to calculate PCI.

The performance comparison of PDB method and ISWP method in light of certain essential aspects is summarized underneath.

- Compared to the PDB method, the ISWP method's performance is better for intricate trend characteristics because it uses few recent data points for prediction rather than the whole dataset. However, if most of the seasonally periodic values are outliers, then the average of the data points will also give an outlier as a corrected value. Further, PCI being confidence coefficient dependent, in the ISWP method, the lower bound of PCI may be negative for specific confidence coefficient values.
- When applied to different polluted datasets originated from the same clean dataset with a fixed number of outliers introduced randomly, the detection performance of the ISWP method is not stable, whereas that for the PDB method is less affected. This is because the portrait dataset formation does not allow the containment of high-density outlier regions even though there is a chance of the same in the original time series. But the ISWP method is directly applied to the actual time series hence is affected by outliers.
- The correction performances of both methods are nearly stable even if the indices of randomly introduced outliers are changed. The ISWP method replaces the outlier for a detected outlier by the average of its seasonally periodic data points and predicted value. In contrast, the PDB method replaces the outlier by the median of that portrait dataset. The index of an outlier does not have any influence over the above correction approaches. Therefore, the correction performance for both methods is stable even after changing the index of the outliers.
- In a nutshell, the ISWP method performs better than the PDB method as the former detects fewer false-positive outliers. But, on the other hand, the PDB method performs better than the ISWP method as it detects a comparatively more number of true-positive outliers.

A preprocessing method's stable detection and correction stages for any outlier introducing mechanisms are always of interest. Testing a method's capability considering several possible outliers introducing mechanisms opens up opportunities to infer the possible reasons for a method's failure to preprocess outliers by particular pollution introducing mechanisms and instigate plans for necessary improvements/hybridizations. The two critical aspects that always make a preprocessing method effective are the accuracy with which outlier regions are constructed during the detection step and how tactically the false outliers are ignored from the correction step.

3.1 Metrics for outlier detection and correction

Calculation of F-score as the harmonic mean of metrics precision (P) and recall (R) is a well-established outlier detection index. The range of the F-score is zero to one. Conventionally, F-score is zero when no true-positive outliers are detected, hence represents the worst-case scenario. Under such conditions, both P and R are zero, resulting F-score of 0/0 as per the F-score formulation. But 0/0 case corresponds to the situation where an ideal method is applied to clean data. Therefore, the lowest value of the F-score is generally set to 0. Mean absolute error (MAE) and root mean square error (RMSE) calculated between clean data and a method's preprocess data are frequently used in the literature[25,27] to evaluate the overall outlier detection and correction performance of methods.

In most of the cases, ISWP method has a higher F-score because its false-positive count is less. But MAE is not lower for the ISWP method than the PDB method because the latter has better correction no matter the FP count. On the other hand, methods with varying counts of FP but ideal correction capability yield similar preprocessed data. So the MAE value remains the same, indicating that the FP count does not affect its calculation. Therefore, the detection performance is underweighted in MAE calculation. Further, for an adequate FP count, any obtained low MAE value is insufficient to guarantee a method's effective preprocessing because the false outliers might have been changed during the correction stage.

4. Probabilistic forecasting using machine learning algorithms

Many ML algorithms being restricted to the training data domain are incapable of extrapolating patterns outside the domain. An algorithm's deriving confidence interval is another property that many ML algorithms do not have, as they are not based on statistical distributions. Besides, assessing how well a forecasting model able to capture patterns is also essential.

Linear regression is the basic ML approach used intensively in the literature for mean prediction.[19] An apt choice of regressor set improves prediction accuracy. Linear regression can be easily extended to QR which is the ever simplest nonparametric probabilistic forecasting approach used to construct prediction intervals using quantiles.[5] Adaption of the pinball loss function (or check function) permits gradient descent-based learning algorithms to yield probabilistic forecasts, and the resulting algorithm is referred to as quantile regression neural network.[28] Tree-based learning algorithms can also be extended for probabilistic forecasting, for example, quantile regression forests (QRF).[21] Few other well-established ML-based probabilistic forecasting methods quantile k-nearest neighbors (QkNN),[29] averaging approaches such as quantile regression averaging (QRA),[30] etc. This chapter thoroughly elucidates the extension of ML and DL methods for probabilistic forecasting in power system applications. Each one of the above methods with a detailed mathematical treatment is elucidated extensively in the literature.[31]

4.1 Forecast evaluation metrics

The use of appropriate metrics is vital to avoid any unbiased comparison. The famous error metrics, such as MAE and RMSE, are used for assessing point forecasting results of several methods. Whereas the quality of probabilistic forecasts can be evaluated using metrics such as quantile probability (QP), quantile score (QS), Winkler score (WS), prediction interval coverage probability (PICP), etc. Formulations of the above point and probabilistic forecasting evaluation metrics are well documented.[31]

QP measures the fraction of data points lying below a specific quantile. It is the mean of all fractions for all quantiles. Comparison using QP is biased as the fraction of data points under a particular quantile is approximately equal to the theoretical quantile value for any obtained smooth quantile forecasts (which is not necessarily a good forecast result). PICP is also similar to QP, but it gives the fraction of points lying between a pair of quantiles (PIs). Since 90% PI is the highest PI generally constructed, the quantiles used are extreme, so there are more chances of data to lie within the PI, thus giving a better PI. But PICP can be artificially improved by widening the PIs, neglecting the interval width and data variance. PICP is an intuitive and easy-to-use metric, so it is frequently used in the literature. However, it is not a very good metric when the interval width (variance) of data is studied. On the contrary, QS weighs the different quantiles using the pinball loss function (whereas QP only gives a fraction of points) and is a better score for comparing quantiles. Similarly, WS considers both coverage and interval width of PIs (whereas PICP neglects interval width). So, QS and WS are, respectively, the better scores than QP and PICP.

5. Performance comparison of well-established probabilistic PV generation forecasting models

5.1 Data preparation

Comparison of a set of simulation results is performed considering historical PV generation samples for 5 years from three locations in the USA (refer to Fig. 10.1) at a daily resolution corresponding to Noon.[32] Missing samples were occasionally present (less than 2% of the entire dataset), updated with the recent previous sample values to restore the time series property. Outliers in the considered time series are detected and corrected using the ISWP method.

5.2 Performance comparison through suitable metrics

Performance of quantile-based models; basic QR, Q*k*NN, QRF, QRA, and Q*k*NNRA are compared using MAE, RMSE, and QS for all the three PV arrays (refer to Table 10.1). For both QRA and Q*k*NNRA models,

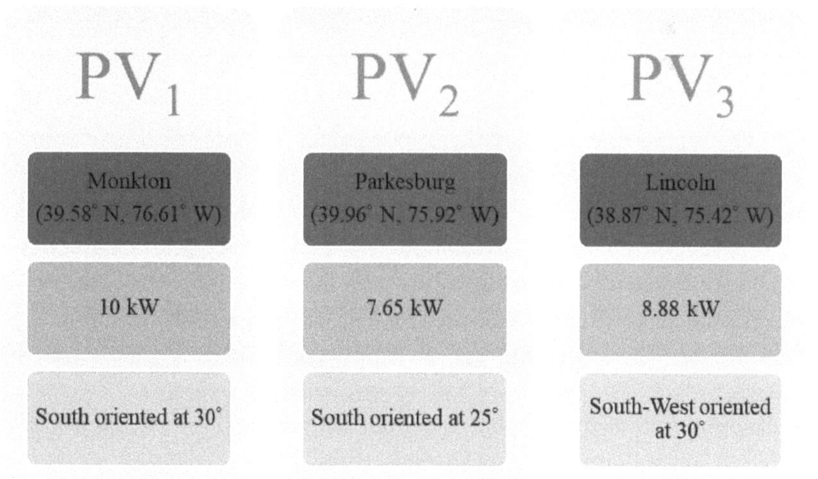

Figure 10.1 Details of PV arrays.

Table 10.1 Comparison of various metrics for different quantile-based models for the three PV arrays at Noon.

	PV_1			PV_2			PV_3		
Model	**MAE**	**RMSE**	**QS**	**MAE**	**RMSE**	**QS**	**MAE**	**RMSE**	**QS**
Basic QR	2.623	3.281	1.477	2.062	2.531	1.127	2.179	2.725	1.052
Q*k*NN	2.576	3.244	1.442	2.022	2.496	1.076	2.041	2.616	0.952
QRF	2.497	3.195	1.475	2.036	2.527	1.145	2.021	2.621	0.888
QRA	2.089	2.546	1.101	1.503	1.820	0.788	1.645	2.025	0.818
Q*k*NNRA	2.012	2.535	1.042	1.473	7.823	0.757	1.547	1.951	0.768

individual point forecasters are ARCH, MLR, and RF. Point forecasters' training period is 3 years from 2012 to 2014, and out-of-sample predictions are made for the remaining 2 years, that is, 2015 and 2016. Regressors for point forecasters MLR and RF and basic QR, QRF, and Q*k*NN are the theoretically formulated ones as used in Ref. 19. For all the five probabilistic forecasting models, 2015 and 2016 are used as training and testing periods, respectively. The optimal value of *k* in Q*k*NN and Q*k*NNRA using the procedure as discussed in Ref. 31. The comparison of various models in Table 10.1 indicates that the quantile forecasts using Q*k*NNRA have superior accuracy than the other models as the obtained error values for Q*k*NNRA are less for all the considered cases. The metric MAE is calculated for the quantiles 0.01, 0.02, ..., 0.99, and the average is noted. Whereas RMSE is calculated considering median forecast results.

6. Conclusion

This chapter highlighted the importance of probabilistic forecasting in the application of power system planning and operation. The importance of data preprocessing is envisioned, and the performance of two outstanding data preprocessing methods is enlightened. Their merits and demerits in handling different effects in the time series are exposed. The mainstream probabilistic forecasting extensions of ML-based point forecasting algorithms are reviewed in detail and are applied to historically collected PV generation data. The result analysis section mainly focused on comparing probabilistic forecasting results using reliable evaluation metrics for PV-integrated power system planning applications. The considered models are equally applicable for obtaining probabilistic forecasts of other power system inputs.

References

1. Morales JM, Conejo AJ, Madsen H, et al. Renewable energy sources—modeling and forecasting. In: *Integrating Renewables in Electricity Markets: Operational Problems*. Boston, MA: Springer US; 2014:15−56. https://doi.org/10.1007/978-1-4614-9411-9_2. Accessed November 9, 2021.
2. Prusty BR, Jena D. A spatiotemporal probabilistic model-based temperature-augmented probabilistic load flow considering PV generations. *Int Trans Electr Energy Syst*. 2019;29(5):e2819. https://doi.org/10.1002/2050-7038.2819.
3. Rajanarayan Prusty B, Jena D, Mishra S, Sood YR, Tomar A. *Uncertainty Modeling Steps for Probabilistic Steady-State Analysis*. 2019:1169−1177. https://doi.org/10.1007/978-981-13-6772-4_102.
4. Lauret P, David M, Pedro HTC. Probabilistic solar forecasting using quantile regression models. *Energies*. 2017;10(10):1591. https://doi.org/10.3390/en10101591.
5. Tripathy DS, Prusty BR, Jena D. Probabilistic forecasting of daily PV generation using quantile regression method. In: *2020 IEEE India Council International Subsections Conference (INDISCON)*. 2020:260−265. https://doi.org/10.1109/INDISCON50162.2020.00060.
6. Wan C, Zhao J, Song Y, Xu Z, Lin J, Hu Z. Photovoltaic and solar power forecasting for smart grid energy management. *CSEE J Power Energy Syst*. 2015;1(4):38−46. https://doi.org/10.17775/CSEEJPES.2015.00046.
7. van der Meer DW, Widén J, Munkhammar J. Review on probabilistic forecasting of photovoltaic power production and electricity consumption. *Renew Sustain Energy Rev*. 2018;81:1484−1512. https://doi.org/10.1016/j.rser.2017.05.212.
8. Prusty BR, Jena D. Combined cumulant and Gaussian mixture approximation for correlated probabilistic load flow studies: a new approach. *CSEE J Power Energy Syst*. 2016;2(2):71−78. https://doi.org/10.17775/CSEEJPES.2016.00024.
9. Prusty BR, Jena D. A critical review on probabilistic load flow studies in uncertainty constrained power systems with photovoltaic generation and a new approach. *Renew Sustain Energy Rev*. 2017;69:1286−1302. https://doi.org/10.1016/j.rser.2016.12.044.
10. Prusty BR, Jena D. An over-limit risk assessment of PV integrated power system using probabilistic load flow based on multi-time instant uncertainty modeling. *Renew Energy*. 2018;116:367−383. https://doi.org/10.1016/j.renene.2017.09.077.
11. Alam AM, Asif Razee I, Zunaed M. Solar PV power forecasting using traditional methods and machine learning techniques. In: *2021 IEEE Kansas Power and Energy Conference (KPEC)*. 2021:1−5. https://doi.org/10.1109/KPEC51835.2021.9446199.
12. Mutavhatsindi T, Sigauke C, Mbuvha R. Forecasting hourly global horizontal solar irradiance in South Africa using machine learning models. *IEEE Access*. 2020;8: 198872−198885. https://doi.org/10.1109/ACCESS.2020.3034690.
13. Rajanarayan Prusty B, Jena D, Mishra S, Sood YR, Tomar A. *Probabilistic Load Flow in a Transmission System Integrated with Photovoltaic Generations*. 2019:1159−1168. https://doi.org/10.1007/978-981-13-6772-4_101.
14. Hor C-L, Watson SJ, Majithia S. *Daily Load Forecasting and Maximum Demand Estimation Using ARIMA and GARCH*. 2006:1−6.
15. Laouafi A, Mordjaoui M, Laouafi F, Boukelia TE. Daily peak electricity demand forecasting based on an adaptive hybrid two-stage methodology. *Int J Electr Power Energy Syst*. 2016;77:136−144. https://doi.org/10.1016/j.ijepes.2015.11.046.
16. Nasmus Sakib Khan Shabbir M, Zawad Ali M, Sifatul Alam Chowdhury M, Liang X. A probabilistic approach for peak load demand forecasting. In: *2018 IEEE Canadian Conference on Electrical Computer Engineering (CCECE)*. 2018:1−4. https://doi.org/10.1109/CCECE.2018.8447641.

17. Leite da Silva AM, Ribeiro SMP, Arienti VL, Allan RN, Do Coutto Filho MB. Probabilistic load flow techniques applied to power system expansion planning. *IEEE Trans Power Syst.* 1990;5(4):1047–1053. https://doi.org/10.1109/59.99351.
18. Fan M, Vittal V, Heydt GT, Ayyanar R. Probabilistic power flow studies for transmission systems with photovoltaic generation using cumulants. *IEEE Trans Power Syst.* 2012;27(4):2251–2261. https://doi.org/10.1109/TPWRS.2012.2190533.
19. Prusty BR, Jena D. Preprocessing of multi-time instant PV generation data. *IEEE Trans Power Syst.* 2018;33(3):3189–3191. https://doi.org/10.1109/TPWRS.2018.2799487.
20. Tripathy DS, Prusty BR, Jena D. Short-term PV generation forecasting using quantile regression averaging. In: *2020 IEEE International Conference on Power Systems Technology (POWERCON).* 2020:1–6. https://doi.org/10.1109/POWERCON48463.2020.9230535.
21. Tripathy DS, Rajanarayan Prusty B, Jena D, Sahu MK. Multi-time instant probabilistic PV generation forecasting using quantile regression forests. In: *2020 IEEE 9th Power India International Conference (PIICON).* 2020:1–6. https://doi.org/10.1109/PIICON49524.2020.9112880.
22. Zhou Y, Sun Y, Wang S, et al. A very short-term probabilistic prediction method of wind speed based on ALASSO-nonlinear quantile regression and integrated criterion. *CSEE J Power Energy Syst.* 2021:1–9. https://doi.org/10.17775/CSEEJPES.2020.05370.
23. Review of preprocessing methods for univariate volatile time-series in power system applications - ScienceDirect. https://www.sciencedirect.com/science/article/abs/pii/S0378779620306830.
24. Chen J, Li W, Lau A, Cao J, Wang K. Automated load curve data cleansing in power systems. *IEEE Trans Smart Grid.* 2010;1(2):213–221. https://doi.org/10.1109/TSG.2010.2053052.
25. Ranjan KG, Tripathy DS, Prusty BR, Jena D. An improved sliding window prediction-based outlier detection and correction for volatile time-series. *Int J Numer Model Electron Network Dev Field.* 2021;34(1):e2816. https://doi.org/10.1002/jnm.2816.
26. Tang G, Wu K, Lei J, Bi Z, Tang J. From landscape to portrait: a new approach for outlier detection in load curve data. *IEEE Trans Smart Grid.* 2014;5(4):1764–1773. https://doi.org/10.1109/TSG.2014.2311415.
27. Ma L, Gu X, Wang B. Correction of outliers in temperature time series based on sliding window prediction in meteorological sensor network. *Information.* 2017;8(2):60. https://doi.org/10.3390/info8020060.
28. Amalia FF, Suhartono, Rahayu SP, Suhermi N. *Quantile Regression Neural Network for Forecasting Inflow and Outflow in Yogyakarta.* Vol. 1028. 2018:012232. https://doi.org/10.1088/1742-6596/1028/1/012232.
29. Bracale A, Carpinelli G, De Falco P. Developing and comparing different strategies for combining probabilistic photovoltaic power forecasts in an ensemble method. *Energies.* 2019;12(6):1011. https://doi.org/10.3390/en12061011.
30. Tripathy DS, Prusty BR. Quantile regression averaging-based probabilistic forecasting of daily ambient temperature. *Int J Numer Model Electron Netw Devices Fields.* 2021;34(3):e2846.
31. Tripathy DS, Prusty BR, Tomar A, Kandari R. Chapter 10 - forecasting of renewable generation for applications in smart grid power systems. In: *Advances in Smart Grid Power System.* Academic Press; 2021:265–298. https://www.sciencedirect.com/science/article/pii/B9780128243374000102. Accessed November 9, 2021.
32. PVOutput. https://www.pvoutput.org/.

CHAPTER ELEVEN

Machine learning application to industrial control systems

Kishore Bingi[1], B. Rajanarayan Prusty[1], Rosdiazli Ibrahim[2]
[1]School of Electrical Engineering, Vellore Institute of Technology, Vellore, Tamil Nadu, India
[2]Department of Electrical and Electronics Engineering, Universiti Teknologi PETRONAS, Seri Iskandar, Malaysia

Contents

1. Introduction

Modeling and controlling the process is quite challenging due to high nonlinearity conditions, sensitivity, and dynamic behavior of the systems. This phenomenon is widespread in all systems, and linearization methods commonly used may not be sufficient in some of the more sensitive applications. Model predictive controllers with linear, nonlinear, fuzzy, and hybrid models have been used to control these types of processes for several years. This assumption is no longer valid for time-varying processes that operate in a broader range of nonlinearity conditions. Nonlinear models, based on gray box models, such as Volterra, Wiener, and Laguerre series

Smart Electrical and Mechanical Systems
ISBN: 978-0-323-90789-7
https://doi.org/10.1016/B978-0-323-90789-7.00008-7

models with uncertain coefficients, are generally split into a nominal and an uncertainty model to permit robust analysis tools. However, the split models will increase the computational load and not guarantee the global optimum to be reached. Black box models, such as fuzzy and neuro-fuzzy models, will not be able to provide any knowledge of the internal structure of the process. It is expected that by using a gray box model, incorporating some physical models into process description, a better understanding of the internal dynamic is predicted, so better control can be designed. Thus, this study aims to model the process's dynamic and nonlinear behavior using neural networks (NNs) to achieve adequate modeling and set-point tracking.

The remaining sections of the chapter are organized as follows: Sections 11.2 and 11.3 demonstrate the application of various modeling and control strategies, including the machine learning techniques for continuous stirred tank reactor (CSTR). Section 11.4 gives the NN-based modeling and control of CSTR. Section 11.5 gives the NN model's performance analysis compared with the state-of-the-art approaches. Section 11.6 concludes the chapter.

2. Modeling strategies of continuous stirred tank reactor

The modeling strategies are classified into three categories as fundamental, empirical, and hybrid. The fundamental modeling is based on understanding the physical properties of the process. In empirical modeling, the model will be developed without applying any process's prior knowledge and utilizing all the model parameters. Hybrid modeling is a combination of both these models. Over the years, numerous models have been reported in the literature using these three strategies will be explained underneath.

2.1 Fundamental modeling approach

Fundamental modeling involves understanding the complete process and developing model equations using the principles[1,2] such as equilibria, electroneutrality, etc. This modeling process requires extensive engineering knowledge to formulate system dynamics and solve them using various mathematical approaches.[3] Although this modeling approach is prevalent, most of the literature's models are complex and based on the operating chemical systems.[4] Therefore, an accurate prediction of system dynamics over broad operational situations is needed in fundamental modeling.

2.2 Empirical modeling approach

In empirical modeling, the first stage is to obtain the system data at various operating conditions. Then, a model will be developed by fitting the generated system data. The three types of fitting models are linear, multiple linear, and nonlinear. The subsequent section will explain the development of these various models as reported in the literature.

2.3 Linear and multiple linear modeling techniques

As reported in Ref. 1, the linear model of the CSTR is developed from the plant's response. In this approach, the fitting of nonlinear data of process plant in a linear design causes many inconsistencies between the original plant and the developed model. Linear models have a simple design and many advantages in developing the controller due to less computational time. However, due to colossal modeling error, linear modeling of the process plant is not adequate in many real-time applications. Thus, the use of multiple linear models is a proper choice for incorporating all the nonlinearities of the process.[5–8]

The authors of Ref. 6 have used multiple linear modeling approaches to incorporate the process's nonlinear behavior. In the first stage of this modeling approach, the entire nonlinear behavior of the process will be obtained and separated by regions. Then, a linear model will be developed at each part with minimal inconsistencies between the original data and the developed model. Finally, a linear combination of these linear models from each area of operation will be designed to form multiple linear models that incorporate the entire process's nonlinear behavior. Furthermore, the authors have proposed a weighting approach during the linear combination of models, and using the modeling error, the weight for each linear model is calculated. The method has proven that the weighted model has shown the best performance on incorporating the complete nonlinear behavior of the process.

Similarly, the Bayesian weighting technique that helps choose the weights for switching between two or more models has been used.[7] These works have proven that the multiple linear modeling approaches have performed better on incorporating the nonlinearities of the process. Furthermore, a linear combination of models using fuzzy logic weighting is proposed,[5,8] which weighs instead of switching. The switching has proven better performance compared to weighting. The best performance is because the weighting approach adds some lag, which degrades the overall performance. Nevertheless, this practice has the drawbacks of high computational time, developing multiple linear models, and selecting a proper switching or weighting.

2.4 Nonlinear modeling techniques

A nonlinear autoregressive exogenous (NARX) identification of the process model has been proposed.[9] This nonlinear modeling approach will always perform better in incorporating the nonlinearities of the process compared to linear and multiple linear modeling approaches. However, the problem with parameter estimation of NARX for high-dimensional systems is unavoidable. In a similar development, Volterra's model with uncertain series coefficients is proposed.[10,11] Volterra's model is an accumulation of various polynomials depicted by a systematic representation in power series expansion. The main advantage of Volterra's approach is that it can easily be divided into uncertainty and nominal models, which helps achieve robust analysis. The authors have also incorporated polynomial chaos expansion for determining the model coefficients.[11] The technique helps in gaining more robust performance in less computational time. However, in some situations, Volterra's model has a high computational load, and due to this, it cannot guarantee the optimal global solution that describes the nonlinearities of the process.

In Wiener and Hammerstein's models, a linear model and a nonlinear model are connected in series.[12] The formation of arrangements for both models is presented in Fig. 11.1. The figure shows that Hammerstein's model is a nonlinear—linear combination, whereas Wiener's model is a linear—nonlinear combination. From the structure, the Hammerstein—Wiener

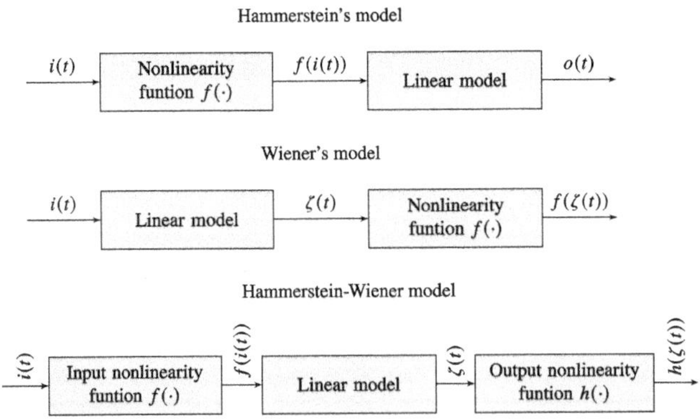

Figure 11.1 Structure of Hammerstein and Wiener's models.

model can be obtained and composed of nonlinear—linear (i.e., Hammerstein model) and linear—nonlinear (i.e., Wiener model) blocks. In the same developments, the authors of Ref. 13 have developed a Wiener—Laguerre model to identify the process. In this model, the linear block has been replaced with Laguerre filters.

Similarly, in Ref. 14, Legendre and Laguerre functions have been used in process identification and illustrated the choice of the models for various applications. The replacement of the nonlinear block of Wiener's model with the piecewise linear function is proposed in Refs. 15 and 16. On the other hand, Wiener—Hammerstein's model performance in identifying processes[12] has shown better performance than multiple linear models. However, the model is so complex and has high computational time.

Takagi and Sugeno proposed a fuzzy modeling approach suitable for the dynamic CSTR processes behavior identification. The main highlight of the fuzzy model is to represent the process model's dynamics in rules. Thus, using the fuzzy model's IF-THEN rules, the input—output relations of a process plant are defined.[17,18]

The fundamental concept of NNs is to create nodes and layers between input and output and link them together. The modeling is achieved using a learning algorithm to obtain proper weights between these nodes and paths. In Refs. 19—21, artificial neural network (ANN) and polynomial models are used for process model identification in which the ANN model of the process outperforms the polynomial ones and obtains better accuracy. Similarly, in Ref. 20, the researchers have developed the process model using an online optimization for weights estimation. In Ref. 22, Elman networks have also been used for process identification. However, extensive training is required to achieve better performance of ANN. The combination of ANN and fuzzy to form neuro-fuzzy modeling techniques has also been proposed by various researchers. In Refs. 23 and 24, a multilayer perceptron-based ANN using a Gath—Geva fuzzy clustering and a LOLIMOT algorithm have been introduced. The identification process combined with a clustering resulted in five operating regions of the process plant. Thus, it helps in developing a linear model and multiple linear models at each functional area. Therefore, the use of ANN has become prevalent in process identification. The popularity is because of the rapid growth in computing power and is easily approachable. However, the most significant concern with the empirical modeling is not knowing the working of the process. Thus, this issue can be tackled using hybrid modeling, which helps incorporate some process models into process specifications.

2.5 Hybrid modeling approach

This modeling approach is a combination of both fundamental and empirical models. In Ref. 25, a hybrid system for modeling the neutralization process plant has been developed. The model is a combination of fundamental and neuro-fuzzy techniques. First, basic modeling equations have been developed using fundamental laws and then by using the backpropagation algorithms for network training. In another work, a hybrid model using ANN and Wiener–Hammerstein model has been proposed.[26]

From the above literature review on modeling of process plant, the comparison between all the three strategies is given in Table 11.1.

3. Controlling strategies of continuous stirred tank reactor

The application of the various model and nonmodel-based control strategies for controlling the process is explained underneath.

3.1 Model-based control strategies

The model-based control strategies include the following:

- Active disturbance rejection control,
- Backstepping,
- Feedback linearization control,
- H-infinity control,
- Model predictive control (MPC), and
- Sliding mode control.

MPC control of nonlinear process requires setting up the controller's specific configuration to achieve good control action of handling the process nonlinearities and disturbances in less computational time. A simple structure of MPC control has achieved better tracking and disturbance rejection performance compared to PID.[1] In comparison, the event-based form of MPC has further performed better compared to simple MPC in handling the process nonlinearities and disturbances.[27,28] This type of structure helps in regulating the process at different operating conditions. However, the process with this adaptive approach is still limited due to its acceptable performance compared to simple structure. Coming to the designs of nonlinear MPC (NMPC), the authors of Ref. 9 have implemented NARX-based NMPC for a water treatment process and compared the performance with linear ones. The version of NMPC has shown similar behavior with the linear MPC.

Table 11.1 Comparing the three modeling methodologies in process identification.

Category	Fundamental	Empirical	Hybrid
Ratio of known and unknown parameters	Low	High	Average
Structure complexity	High	Low	Average
Fundamental contents	Yes	No	Yes/no
Static nonlinearities	Yes	Yes	Yes
Dynamic nonlinearities	Yes	Yes	Yes
Online estimation	No	Yes	Yes
System identification	Single and multicriteria optimization	Standard prediction error methods	Second-order methods
Use of neural networks and fuzzy sets	No	Yes	Yes
Partial differential equation models	Yes	No	No
Application area	System understanding, diagnostics and training	Model-based control, adaptive filtering	Fault detection, process tracking

The robust configurations of NMPC have been designed for Volterra and Wiener's process models.[10,11,13] The model's robustness has been tested using a structured singular value norm. Similarly, Wiener, Hammerstein, and Wiener–Laguerre's process model has also been controlled using robust NMPC and MPC control strategies.[12,13,26] The recent applications of MPC are with using NNs. A linear MPC incorporated with NNs has been introduced in Ref. 25. In the related development, the use of NNs for both process identification and online optimization has been achieved.[21,29] The authors have completed the solving of nonlinear online optimization effectively in less computational time. Implementing neuro-fuzzy modeling for MPC control of process has shown better performance than PID.[23,24] In all these works, the results of NN-based MPC have shown an improvement in the standard forms. In summary, the approaches still lack incorporating robustness and time delays of process, mainly while using nonlinear modeling approaches in the optimization and prediction of MPC.

In two stages, the formulation of the control problem using H-infinity has been achieved. First, represent the problem as a mathematical optimization approach. Then, solve the optimization using a controller. In Ref. 3, the fundamental modeling for the process and H-infinity control have been designed and compared with the model reference adaptive controller. Similarly, in Ref. 30, the modeling and controlling of the process are achieved using fuzzy and a robust H-infinity controller, respectively. However, the H-infinity has significant limitations, such as the requirement of a high mathematical understanding level for applying to a process plant that is to be controlled.

In Ref. 2, the ARDC of the neutralization process has been studied. This work uses the fundamental modeling approach based on physical and chemical properties for process model development. Furthermore, the work defines a one-to-one mapping control variable as the existing variable pH is an affine control parameter. A tuning method has also been proposed for achieving better control parameters. Similarly, in the feedback linearization approach, the first step is transforming the nonlinear system into a linear one using process variables and an appropriate control input. Then, linear control techniques have to be applied. In Ref. 31, this approach has been used to transform the developed fundamental process model into a linear one using the pole placement method. However, the simulation study showed a high steady-state error.

The structure of the backstepping and global linearization controllers is similar. The adaptive global linearization control has consistently shown better performance than backstepping. In Ref. 31, an adaptive backstepping controller has been designed for process plants similar to the global linearization control structure. In both cases, the controllers are developed on a reduced-order model for easy design of state estimator. In Ref. 32, for the control pH value in the reactor tank, a fuzzy sliding mode controller is used.

Similarly, in Ref. 33, the same type of control strategy is developed to control pH and liquid level in the CSTR tank. Both works' fuzzy and sliding mode combination has shown better performance than using two single fuzzy controllers. However, the simulation results show a chattering problem in the fuzzy approaches. The backstepping controller results showed that the conventional version of global linearization control has better set-point tracking performance but failed to keep the system stable under disturbances.

3.2 Nonmodel-based control strategies

Some widely used nonmodel-based control strategies include PID, fractional PID, fuzzy, and NN controllers. Most industrial process control uses PID controllers owing to their simple design and accessible controller parameters tuning methods.[34,35] While designing the PID, the most critical task is the proper selection of parameters. Ziegler–Nichols technique is probably the most widely accepted approach for controller parameter tuning. Other methods used for parameter tuning include the Branica method, Cohen–Coon method, Marlin method, Smith method, and Tyreus–Luyben method.[36,37] From these classical techniques, it is worth noting that all the techniques require high numerical computations for proper parameter selection. Therefore, researchers have started using optimization algorithms for parameter tuning. The wide use of optimization algorithms is because of comprehensive developments in this area and more possibilities of achieving better performance. The most recent heuristic methods that are used for tuning PID are ant colony, genetic algorithm, and particle swarm.[38–41] However, PID controllers are quite inadequate during load disturbances. Set-point change and fixed gains of PID fail to keep the set-point characteristics within the acceptable limit. On the other hand, the PID controller's fixed gain parameters fail to reduce the oscillations in output response.[42,43] The PID controller's performance can be enhanced using fractional I and D actions. In Ref. 44, FOPID is designed for the fundamental model of pH neutralization plant. Similarly, in Ref. 45, FOPID is intended for Wiener's model of pH plants. Here, the Wiener model's linear part is identified as FOPDT, and the model's nonlinear part is specified using the fuzzy technique. Furthermore, 2 DoF fractional PIDs are designed for process plants' fundamental and practical models.[41,46] However, the simulation results show that a single FOPID controller cannot control the pH process with good performance in all set points.

A fuzzy logic control strategy has been developed for the process plant, and the controller's performance is compared with the PID. This controller's design analysis uses process response in developing a first-order plus dead time model. Similarly, fuzzy logic and Lyapunov adaptive control schemes have been designed for the neutralization process. However, the developed method uses Mamdani and Takagi–Sugeno fuzzy inference and is optimized using an optimization algorithm.[25] In Ref. 47, genetic optimized adaptive fuzzy is designed for the neutralization process. In these works, optimization algorithms optimize the membership functions of fuzzy inference.[48] The PID performance has also been enhanced by combining with fuzzy as it can deal with nonlinear systems effectively. The fuzzy PID is more appropriate for controlling the nonlinearities system more efficiently. In Refs. 44,49, fuzzy logic has been used for PID parameter tuning. In the design, the developed process model is based on fundamental concepts, and the performance of fuzzy PID and conventional PIDs have been compared. In Refs. 50,51 an online tuning of fuzzy PID to tune the two input scaling factors and integral and derivative components of PID has been proposed. In Ref. 47, self-organizing is developed for modifying the output membership functions using a genetic algorithm.

Applying the NN-based approach of MPC control is another standard and straightforward implementation to the nonlinear process plant. The NN-based MPC has been used as a forward process model for output prediction. In Ref. 22, Elman networks have been utilized for training the experimental dataset, and in Refs. 20,21, multilayered NNs have been used. In these works, the designed controller is a constrained optimization-based technique for predicting process outputs. In a parallel development, the authors of Ref. 52 have combined NNs and a fundamental approach for real-time optimization and MPC control of CSTR. The authors of Ref. 53 have developed an MPC controller for CSTR using deep learning techniques. These works summarized that there is still some scope to enhance the performance in the control of CSTR by utilizing NNs.

4. Neural network–based modeling and control of CSTR

4.1 Process modeling

The benchmark model of CSTR is shown in Fig. 11.2. As shown in the figure, the objective of this process is to maintain the value of output C_o

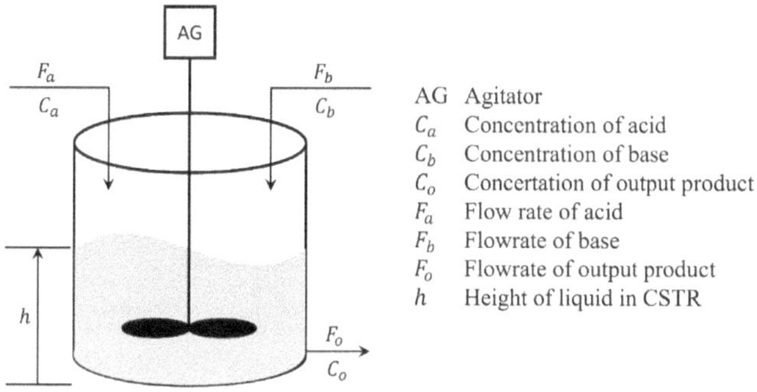

Figure 11.2 Benchmark model of continuous stirred tank reactor (CSTR).

constantly. The constant value has to be achieved by varying the input flow rates F_a and F_b. However, in this work, for simplicity in designing the controller, only flow rate F_a is varied by keeping the constant flow rate F_b and $h(t)$ is uncontrolled. The dynamic model of CSTR is given as follows:[41,54]

$$\frac{d}{dt}h(t) = F_a + F_b - 0.2\sqrt{h(t)}, \tag{11.1}$$

$$\frac{d}{dt}C_o(t) = (C_a - C_o(t))\frac{F_a(t)}{h(t)} + (C_b - C_o(t))\frac{F_b(t)}{h(t)} - \frac{C_o(t)}{(1 + C_o(t))^2}. \tag{11.2}$$

The above model is highly nonlinear and sensitive. Thus, by simulating this dynamic model, a dataset has been created. The parameters used for the simulation are $C_a = 25$ mol/L, $C_b = 0.1$ mol/L, and $F_b = 0.1\ \text{m}^3/\text{h}$. The dataset contains 16,000 samples, which have been generated by randomly varying the input $F_a \in (0, 5)\ \text{m}^3/\text{h}$ and the output C_o is computed. The generated dataset will be utilized for developing the process of NN model. Fig. 11.3 shows the block diagrammatic representation of process modeling using NNs. The block diagram shows that the network model arranges to capture and represent the process plant's forward dynamics. In the figure, $u(t)$ and $y_p(t)$ denote the plant's input and output, $y_m(t)$ is the NN's model output, and $e(t)$ is the error between $y_p(t)$ and $y_m(t)$. This error signal is used during the process of network's training as a training signal.

Fig. 11.4 presents the architecture of NN model. The architecture shows that the first, middle, and last layers as input, hidden, and output. The network uses previous values of plant's input $(u(t-1),\ u(t-1))$ and output

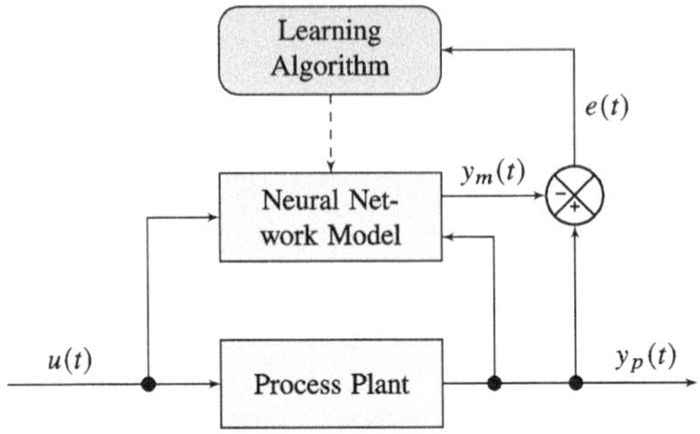

Figure 11.3 Block diagrammatic representation of process modeling using neural networks.

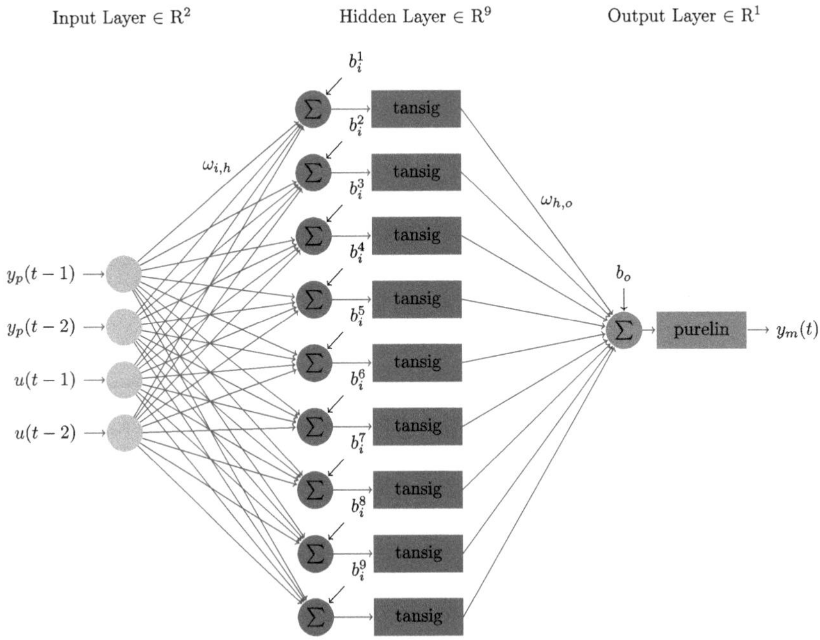

Figure 11.4 Architecture of neural networks model.

$\left(y_p(t-1),\ y_p(t-2)\right)$ to predict the future value of model output $y_m(t)$. Thus, input layer has four nodes representing $u(\cdot)$ and $y_p(\cdot)$. The output layer has a single node representing $y_m(t)$. The number of hidden layer nodes is computed as follows:[55,56]

$$N_h = \frac{4.5 \times N_i}{N_i - 1}, \tag{11.3}$$

where "N_i" and "N_h" are the number of input and hidden nodes, respectively.

For training the network model, the Levenberg–Marquardt algorithm is utilized in this work. This hybrid algorithm achieves robust optimal solutions by using both Gauss–Newton and gradient descent approaches. According to the learning algorithm, the weight update rule for getting the new weights ω_{c+1} using the current weight ω_c and cumulative error e is given as[57]

$$\omega_{c+1} = \omega_c - \left[J^T J - l_p I\right]^{-1} J^T e, \tag{11.4}$$

where I and J are the identity and Jacobian matrices, and l_p is the learning parameter.

The architecture in Fig. 11.4 also shows that the selected activation function at the hidden is *tansig* and at the output is *purelin*. The mathematical formulations of both of these functions are given as[58]

$$\phi(x) = \frac{2}{1 + \exp(-2x)} - 1, \tag{11.5}$$

$$\phi(y) = y. \tag{11.6}$$

The above combination (nonlinear–linear) of functions is usual in literature and proven for achieving an effective training process. The delta rule for updating the weights $\omega_{i,h}$ (between input and hidden nodes) and $\omega_{h,o}$ (between hidden and output nodes) is computed as follows:[59]

$$\omega_{i,h} \leftarrow \omega_{i,h} + l_r \delta_h x_i, \tag{11.7}$$

$$\omega_{h,o} \leftarrow \omega_{h,o} + l_r \delta_o y_h. \tag{11.8}$$

In the above equations,

x_i and y_h are the current values of the input and hidden nodes,

ranges of i and h are $[1, 2, \ldots, N_{i-1}, N_i]$ and $[1, 2, \ldots, N_{h-1}, N_h]$, respectively, and

$l_r \in (0, 1)$ is the learning rate.

Furthermore, the formula for computing the values of δ_h and δ_o is given as

$$\delta_h = \phi'(v_h) e_h, \delta_o = \phi'(v_o) e_o, \tag{11.9}$$

where

e_h and e_o are the errors at j^{th} hidden and o^{th} output nodes,

v_h and v_o are the weighted sum of inputs and hidden nodes along with the bias b_i and b_o, respectively, and

$\phi'(\cdot)$ and $\phi'(\cdot)$ are the first derivatives of the activation functions.

For measuring the model's prediction strength and performance, the R^2 and mean square error (MSE) are adopted in this work. The formula for calculating these measures for the number of samples "n_s" is as follows:[56,57]

$$\mathrm{R}^2 = 1 - \frac{\sum_{k=1}^{n_s}\left(Y_{A,k} - Y_{P,k}\right)^2}{\sum_{k=1}^{n_s}\left(Y_{A,k} - Y_{Avg}\right)^2}, \tag{11.10}$$

$$\mathrm{MSE} = \frac{1}{n_s}\sum_{k=1}^{n_s}\left(Y_{A,k} - Y_{P,k}\right)^2, \tag{11.11}$$

where Y_A and Y_P denote the actual and predicted data, and Y_{Avg} represents the average values of Y_A.

The value of $R^2 \in (0, 1)$ indicates the networks' prediction strength. The closing value of R^2 to one ensures the best fit of the proposed network. Similarly, the MSE value measures the error between Y_A and Y_P. Thus, a closer value of MSE to zero represents the prediction network's better estimation.

4.2 Process controlling

Fig. 11.5 shows the block diagrammatic representation of process controlling using the trained NN model (refer to Fig. 11.3). In the block diagram, y_r is the reference, N_c and N_1 are the control and prediction horizons, and N_p is the number of previous values used in simulation. As shown in the block diagram, the referent input y_r minimizes the cost function block.

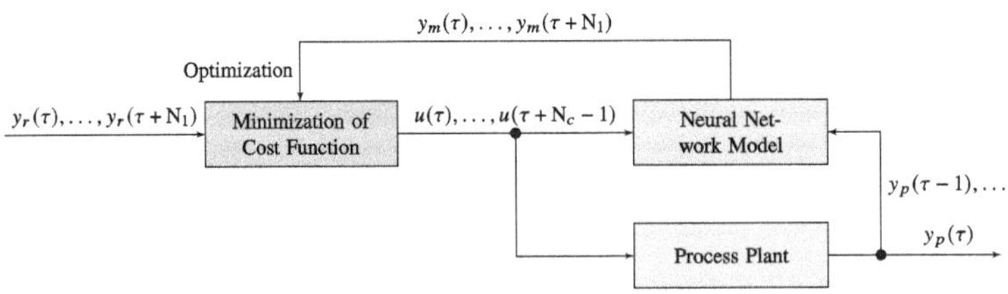

Figure 11.5 Block diagrammatic representation of process controlling using neural networks.

The result of this block is the control signal u, which will be given to the NN model and the process plant. The cost function used in this work is a combination of MSE between y_r and y_m, and weighted square of the change in control u is given as follows:

$$J_c = \sum_{i=N_1}^{N_2} (y_r(\tau + i) - y_m(\tau + i))^2 + \lambda \sum_{i=1}^{N_c} (u(\tau + i) - u(\tau + i))^2, \tag{11.12}$$

where λ is the weighting factor that balances between the two factors (i.e., MSE between y_r and y_m and weighted square of the change in control u).

As seen from J_c, the plants predictions will run from future time stamps N_1 to N_2. The control horizon bound is N_c. It is to note that the constraints on time stamps are the values of $N_1, N_c \leq N_2$. Also, the weight factor λ will act as a damper in predicting u.

5. Results and discussions

In this section, the NN model's performance analysis for both process modeling and control of CSTR has been analyzed. Furthermore, the performance is compared with the nonlinear autoregressive moving average (NARMA) model. As mentioned earlier, the dataset consists of 16,000 samples which have been generated by randomly varying the input $F_a \in (0, 5)\ \mathrm{m}^3/\mathrm{h}$ and the output C_o is computed. Thus, the generated input F_a and output C_o are shown in Fig. 11.6. Among the generated 16,000 samples of data, 12,000 and 4000 samples have been used for training and testing, respectively.

For training the NN model in Fig. 11.3 and Fig. 11.4, the training algorithm used is the Levenberg–Marquardt technique, and the considered activation functions at hidden and output layers are *tansig* and *purelin*, respectively. As seen from Fig. 11.4, with four inputs $\left(y_p(t-1), y_p(t-2), u(t-1), u(t-2)\right)$, the number of hidden layer nodes using the thump rule is estimated as 6. Furthermore, for the NARMA model, same configuration of NN model has been chosen for effective comparison.

The training and testing performance of the NN model is shown in Fig. 11.7. Similarly, the training and testing performance of the NARMA model is demonstrated in Fig. 11.8. Furthermore, the model's prediction

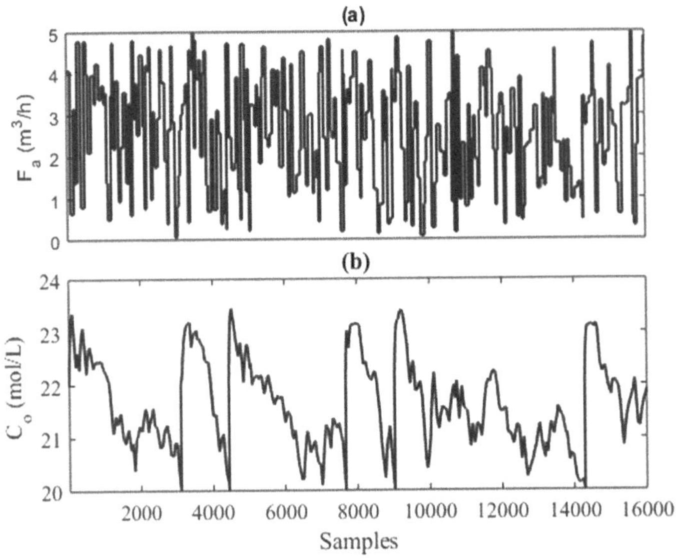

Figure 11.6 Generated input and output data of continuous stirred tank reactor (A) flow rate of acid (B) concentration of output product.

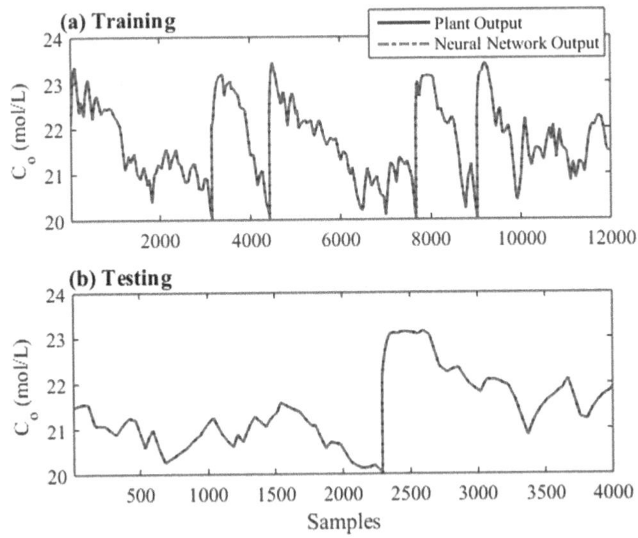

Figure 11.7 Training and testing performance of neural network model (A) training (B) testing.

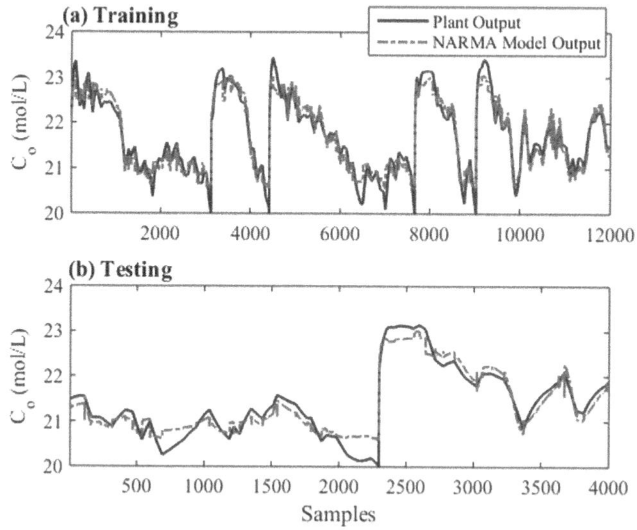

Figure 11.8 Training and testing performance of nonlinear autoregressive moving average model (A) training (B) testing.

Table 11.2 Numerical analysis of neural network (NN) and nonlinear autoregressive moving average (NARMA) models.

	Training		Testing	
Model	**R^2**	**MSE**	**R^2**	**MSE**
NN model	0.9997	2.5160×10^{-7}	0.9998	5.6820×10^{-8}
NARMA model	0.9689	0.0409	0.9631	0.0466

strength and performance in terms of R^2 and MSE are given in Table 11.2. The NN model's tracking performance has shown to be better compared to the NARMA model. The R^2 values of the NN model for training and testing are 0.9997 and 0.9998, respectively, and confirm the NN model's best performance compared to 0.9689 and 0.9631 of NAMA models. Similarly, the MSE values of the NN model for training and testing are 2.5160×10^{-7} and 5.6820×10^{-8}, respectively, and further confirm the NN model best performance compared to 0.0409 and 0.0466 of NAMA models.

For controlling the process with the trained NN and NARMA models, the considered simulation parameters are $N_1 = 1, N_2 = 7, N_c = 2$, and $\lambda = 0.05$. The tracking performance of both the trained models is shown in Fig. 11.9. The numerical analysis for rising, peak, settling times, and

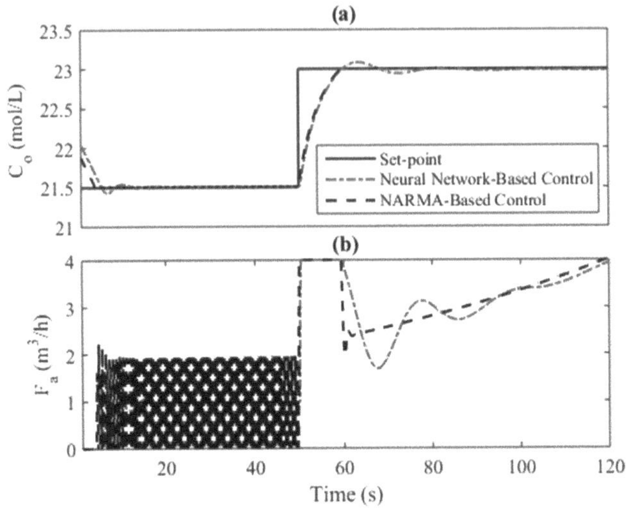

Figure 11.9 Tracking performance of neural network and nonlinear autoregressive moving average–based process controlling (A) concentration of output product (B) flow rate of acid.

Table 11.3 Numerical analysis of neural network (NN) and nonlinear autoregressive moving average (NARMA)–based process controlling.

Approach	t_r	t_p	t_s	%OS
NN-based control	5.2118	63.6000	134.2492	0.7367
NARMA-based control	1.6661	60.0000	197.9297	2.6221

overshoot is tabulated in Table 11.3. The tracking performance of NN-based control has shown better compared to NARMA-based control. The numerical analysis has also shown that the NN-based control has the least overshoot of 0.7367 compared to 2.6221 of NARMA-based control. Also, the NN-based control has the fastest settling time of 134.2492 s compared to 197.9297 s of NARMA-based control. However, the NN-based control has the highest rising and peak times of 5.2118 and 63.6 s compared to 1.6661 and 60 s of NARMA-based control. Furthermore, the NN-based control has also achieved smoother action compared to NARMA-based control. The chattering effect in both the process output and control action can be clearly seen in Fig. 11.10 (zoomed-in version of Fig. 11.9). The chattering will also increase the deterioration rate of process control values.

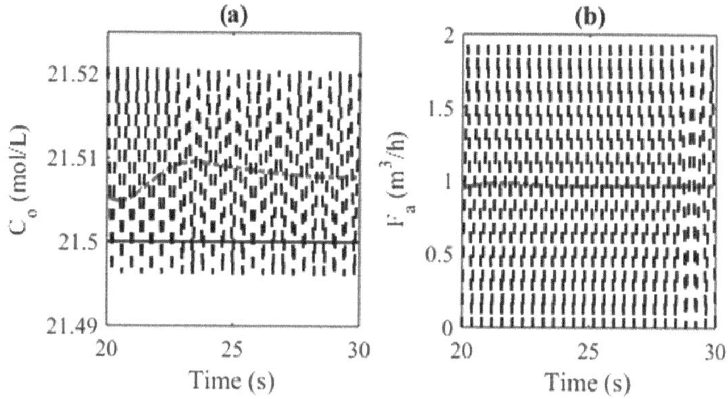

Figure 11.10 Tracking performance of neural network and nonlinear autoregressive moving average–based process controlling (zoomed-in version) (A) concentration of output product (B) flow rate of acid.

6. Conclusion

The growth, development, and future of various machine learning techniques in industrial process control applications are discussed in this chapter. Significant developments in modeling and controlling process strategies are highlighted, but most of the chapter deals with NN-based process modeling and control techniques. The mainstream NN-based modeling and controlling of the process are given importance. The most widely used network models are elaborated, and NN-based modeling and handling of the process are clearly explained along with the simulation study compared to the NARMA-based approach. The study confirms the NN model's best performance compared to NARMA for both process modeling and control.

References

1. Annal WP, Jerome J. Performance assessment of PID and MPC control algorithm subject to servo tracking and disturbance rejection. *Aust J Basic Appl Sci*. 2014;9.
2. Brown A, Zhang J. Active disturbance rejection control of a neutralisation process. In: *Computer Aided Chemical Engineering*. Vol. 33. Elsevier; 2014:739–744.
3. Benny A, Tharakan LG, Jaffar NE, Jaleel JA. *Adaptive and H∞ Control of a Strong Acid—Strong Base System*. 2013:662–667.
4. Elameen MHE, Karsiti MN, Ibrahim R. *Nonlinear Model Feedback Linearization Control Strategy of a pH Neutralization Process*. 2014:1–5.
5. Bello O, Hamam Y, Djouani K. Control of a coagulation chemical dosing unit for water treatment plants using MMPC based on fuzzy weighting. *J Water Proc Eng*. 2014;4:34–46.
6. Costa TV, Fileti AM, Oliveira-Lopes LC, Silva FV. Experimental assessment and design of multiple model predictive control based on local model networks for industrial processes. *Evol Syst*. 2015;6(4):243–253.

7. Hermansson AW, Syafiiey S, Noorz SM. *Multiple Model Predictive Control of Nonlinear pH Neutralization System*. 2010:301–304.
8. Shamsaddinlou A, Fatehi A, Sedigh AK, Karimi MM. *Study of Multiple Model Predictive Control on a pH Neutralization Plant*. 2013:1–6.
9. Bello O, Hamam Y, Djouani K. Nonlinear model predictive control of a coagulation chemical dosing unit for water treatment plants. *IFAC Proc Vol*. 2014;47(3):370–376.
10. Díaz-Mendoza R, Budman H. Structured singular valued based robust nonlinear model predictive controller using Volterra series models. *J Process Control*. 2010;20(5): 653–663.
11. Kumar D, Budman H. Robust nonlinear MPC based on Volterra series and polynomial chaos expansions. *J Process Control*. 2014;24(1):304–317.
12. Khani F, Haeri M. Robust model predictive control of nonlinear processes represented by Wiener or Hammerstein models. *Chem Eng Sci*. 2015;129:223–231.
13. *Wiener Model Identification and Nonlinear Model Predictive Control of a pH Neutralization Process Based on Laguerre Filters and Least Squares Support Vector Machines*. https://link.springer.com/article/10.1631/jzus.C0910779. Accessed 3 October 2021.
14. Dubravić A, Šehić Z, Burgić M. Orthonormal functions based model predictive control of pH neutralization process. *Teh Vjesn*. 2014;21(6):1249–1253. https://hrcak.srce.hr/index.php?id_clanak_jezik=193982&show=clanak. Accessed October 3, 2021.
15. Oblak S, Škrjanc I. Continuous-time Wiener-model predictive control of a pH process based on a PWL approximation. *Chem Eng Sci*. 2010;65(5):1720–1728. https://doi.org/10.1016/j.ces.2009.11.008.
16. Ipanaqué W, Manrique J. Identification and control of pH using optimal piecewise linear Wiener model. *IFAC Proc Vol*. 2011;44(1):12301–12306.
17. Kumbasar T, Eksin I, Guzelkaya M, Yesil E. Interval type-2 fuzzy inverse controller design in nonlinear IMC structure. *Eng Appl Artif Intell*. 2011;24(6):996–1005.
18. Kumbasar T, Eksin I, Guzelkaya M, Yesil E. Type-2 fuzzy model based controller design for neutralization processes. *ISA Trans*. 2012;51(2):277–287.
19. Chaber P, Ławryńczuk M. *Recurrent Polynomial and Neural Structures in Modelling of a Neutralisation Process*. 2015:23–32.
20. Tharakan LG, Benny A, Jaffar NE, Jaleel JA. *Neural Network Based pH Control of a Weak Acid—Strong Base System*. 2013:674–679.
21. Wior I, Boonto S, Abbas HS, Werner H. *Modeling and Control of an Experimental pH Neutralization Plant Using Neural Networks Based Approximate Predictive Control*. International Program Committee; 2010:13.
22. Wysocki A, Ławryńczuk M. *Predictive Control of a Multivariable Neutralisation Process Using Elman Neural Networks*. 2015:335–344.
23. Fatehi A, Sadeghpour B, Labibi B. Nonlinear system identification in frequent and infrequent operating points for nonlinear model predictive control. *Inf Technol Control*. 2013;42(1):67–76.
24. Saadat A, Alvanagh AA, Rezaei H. *pH Control in Biological Process Using MMPC Based on Neuro-Fuzzy Model by LOLIMOT Algorithm*. 2013:1–6.
25. Zanil MF, Norhuda AM, Hussain MA, Omar R. Hybrid model of pH neutralization for a pilot plant. *J Intell Fuzzy Syst*. 2014;26(2):551–561.
26. Ławryńczuk M. Suboptimal nonlinear predictive control based on multivariable neural Hammerstein models. *Appl Intell*. 2010;32(2):173–192.
27. Pawlowski A, Fernández I, Guzmán JL, Berenguel M, Acién FG, Normey-Rico JE. Event-based predictive control of pH in tubular photobioreactors. *Comput Chem Eng*. 2014;65:28–39. https://doi.org/10.1016/j.compchemeng.2014.03.001.
28. Pawlowski A, Mendoza JL, Guzmán JL, Berenguel M, Acién FG, Dormido S. Effective utilization of flue gases in raceway reactor with event-based pH control for microalgae culture. *Bioresour Technol*. 2014;170:1–9.

29. Tang W, Karim MN. Multi-model MPC for nonlinear systems: case study of a complex pH neutralization process. In: *Computer Aided Chemical Engineering*. Vol. 29. Elsevier; 2011:622–627.
30. Dolatabadi MH, Monfared MN, Fakharian A. *Robust H-Infinity Control for the pH Neutralization Process Based on Fuzzy Models*. 2013:1–6.
31. Nejati A, Shahrokhi M, Mehrabani A. Comparison between backstepping and input–output linearization techniques for pH process control. *J Process Control*. 2012;22(1): 263–271.
32. Chen J, Liu Z, Peng Y, Guo Q. *A New Control Method in pH Neutralization Process*. 2013: 511–519.
33. Chen J, Peng Y, Han W, Guo M. Adaptive fuzzy sliding mode control in PH neutralization process. *Proc Eng*. 2011;15:954–958.
34. Chan LLT, Chen T, Chen J. PID based nonlinear processes control model uncertainty improvement by using Gaussian process model. *J Process Control*. 2016;42:77–89.
35. Chien LJ, Ibrahim R, Hassan SM, Bingi K, Supramaniam T. *Comparison between Wired and WirelessHART Networked PID Control of Process Flow. 2018 International Conference on Intelligent and Advanced System (ICIAS)*. IEEE; 2018:1–5.
36. Devan PAM, Hussin FA, Ibrahim R, Bingi K, Abdulrab H. *Fractional-order Predictive PI Controller for Process Plants with Deadtime. 2020 IEEE 8th R10 Humanitarian Technology Conference (R10-HTC)*. IEEE; 2020:1–6.
37. Mudi RK, Dey C. Performance improvement of PI controllers through dynamic set-point weighting. *ISA Trans*. 2011;50(2):220–230.
38. Bingi K, Ibrahim R, Karsiti MN, Hassan SM. *Fractional-order PI-PD Control of Real-Time Pressure Process. Progress in Fractional Differentiation and Applications*. 2020;6(4):289–299.
39. Bingi K, Ibrahim R, Karsiti MN, Chung TD, Hassan SM. *Optimal PID Control of pH Neutralization Plant. 2016 2nd IEEE International Symposium on Robotics and Manufacturing Automation (ROMA)*. IEEE; 2016:1–6.
40. Bingi K, Ibrahim R, Karsiti MN, Hassan SM, Harindran VR. A comparative study of 2DOF PID and 2DOF fractional order PID controllers on a class of unstable systems. *Arch Control Sci*. 2018;28. https://doi.org/10.24425/acs.2018.125487.
41. Bingi K, Ibrahim R, Karsiti MN, Hassan SM, Harindran VR. *Fractional-Order Systems and PID Controllers* 264. Cham: Springer; 2020.
42. Veselý V, Ilka A. Gain-scheduled PID controller design. *J Process Control*. 2013;23(8): 1141–1148. https://doi.org/10.1016/j.jprocont.2013.07.002.
43. Veselý V, Ilka A. Design of robust gain-scheduled PI controllers. *J Franklin Inst*. 2015; 352(4):1476–1494. https://doi.org/10.1016/j.jfranklin.2015.01.009.
44. Bingi K, Ibrahim R, Karsiti MN, Hassan SM. *Fuzzy Gain Scheduled Set-point Weighted PID Controller for Unstable CSTR Systems. 2017 IEEE International Conference on Signal and Image Processing Applications (ICSIPA)*. IEEE; 2017:289–293.
45. Azarmi R, Tavakoli-Kakhki M, MonirVaghefi H, Shaghaghi D, Fatehi A. *Fractional Order Control of pH Neutralization Process Based on Fuzzy Inverse Model*. 2015:817–822.
46. Bingi K, Ibrahim R, Karsiti MN, Hassan SM. Fractional order set-point weighted PID controller for pH neutralization process using accelerated PSO algorithm. *Arabian J Sci Eng*. 2018;43(6):2687–2701. https://doi.org/10.1007/s13369-017-2740-7.
47. Singh PK, Bhanot S, Mohanta HK. Optimized and self-organized fuzzy logic controller for pH Neutralization process. *Int J Intell Syst Appl*. 2013;5(12):99.
48. Singh PK, Bhanot S, Mohanta HK. Differential evolution based optimal fuzzy logic control of pH neutralization process. In: *International Conference for Convergence for Technology-2014*. 2014:1–5. https://doi.org/10.1109/I2CT.2014.7092060.
49. Esfandyari M, Fanaei MA, Zohreie H. Adaptive fuzzy tuning of PID controllers. *Neural Comput Appl*. 2013;23(1):19–28.

50. Heredia-Molinero MC, Sanchez-Prieto J, Briongos JV, Palancar MC. Feedback PID-like fuzzy controller for pH regulatory control near the equivalence point. *J Process Control.* 2014;24(7):1023−1037.
51. Karasakal O, Guzelkaya M, Eksin I, Yesil E, Kumbasar T. Online tuning of fuzzy PID controllers via rule weighing based on normalized acceleration. *Eng Appl Artif Intell.* 2013;26(1):184−197.
52. Zhang Z, Wu Z, Rincon D, Christofides PD. Real-time optimization and control of nonlinear processes using machine learning. *Mathematics.* 2019;7(10):890.
53. Wang G, Jia Q-S, Qiao J, Bi J, Zhou M. Deep learning-based model predictive control for continuous stirred-tank reactor system. *IEEE Transact Neural Networks Learn Syst.* 2021;32(8):3643−3652. https://doi.org/10.1109/TNNLS.2020.3015869.
54. Seborg DE, Mellichamp DA, Edgar TF. *III FJD. Process Dynamics and Control.* John Wiley & Sons; 2010:547.
55. *Review on Methods to Fix Number of Hidden Neurons in Neural Networks.* https://www.hindawi.com/journals/mpe/2013/425740/. Accessed 3 October 2021.
56. Bingi K, Prusty BR, Kumra A, Chawla A. Torque and temperature prediction for permanent magnet synchronous motor using neural networks. In: *2020 3rd International Conference on Energy, Power and Environment: Towards Clean Energy Technologies.* 2021: 1−6. https://doi.org/10.1109/ICEPE50861.2021.9404536.
57. Bingi K, Prusty BR. Chaotic time series prediction model for fractional-order Duffing's oscillator. In: *2021 8th International Conference on Smart Computing and Communications (ICSCC).* 2021:357−361. https://doi.org/10.1109/ICSCC51209.2021.9528128.
58. Chaudhary J, Puri H, Mantri R, Raghavendra KR, Bingi K. Prediction of turbidity in beach waves using nonlinear autoregressive neural networks. In: *2021 8th International Conference on Smart Computing and Communications (ICSCC).* 2021:42−46. https://doi.org/10.1109/ICSCC51209.2021.9528261.
59. Mantri R, Raghavendra KR, Puri H, Chaudhary J, Bingi K. Weather prediction and classification using neural networks and k-nearest neighbors. In: *2021 8th International Conference on Smart Computing and Communications (ICSCC).* 2021:263−268. https://doi.org/10.1109/ICSCC51209.2021.9528115.

CHAPTER TWELVE

Data-driven based optimal power system operation and control in a weak grid

Parvaiz Ahmad, S.A. Lone and Neeraj Gupta
Department of Electrical Engineering, National Institute of Technology, Srinagar, Jammu and Kashmir, India

Contents

Nomenclature

DER Distributed energy resources
DG Distributed generators
EES Energy storage system
ESS Energy storage system
MG Microgrid
MGCC Microgrid centralized controller
ML Machine Learning
NMG Networked microgrid

Smart Electrical and Mechanical Systems
ISBN: 978-0-323-90789-7
https://doi.org/10.1016/B978-0-323-90789-7.00007-5

PCC Point of common coupling
PMU Phasor measurement unit
PV Photo voltaic
RDG Renewable distributed generation
RE Renewable energy
RES Renewable energy sources
SMES Superconducting magnetic energy storage

1. Introduction

In modern power systems, the renewable energy penetration is inTable 12.1 creasing and becoming superior to traditional power generation system throughout the world as currently, conventional power systems are overburdened, have huge transmission and distribution losses and are not in a position to meet all requirements of the consumers. Besides this, traditional power systems rely on thermal power plants and almost 60% of our generation comes from these so-called thermal power plants which incur huge expenses and create a lot of environmental issues and impact the life sustaining environment. Since in our country India, we do not have super thermal power plants.[1] The modern solution for fulfilling power needs and shielding our environment from such growing carbon emissions are distributed generators incorporating renewable energy sources (solar, wind, biomass, etc.), ESSs with intelligent control makes an alternate solution for centralized conventional power system network. These DGRs have technical economic and environmental benefits, can operate in a

Table 12.1 Applications of data-driven techniques in renewable penetrated power system.

S. No.	At generation control center	At transmission level	At distribution operation end	At user end
1	Generation planning	Economic load dispatch	Real time operation	Demand response
2	Economic load dispatch	State estimation	Equipment health monitoring	Load forecasting
3	–	Stability enhancement	Forecasting	Real-time energy management
4	–	Real time monitoring	Planning	Customer behavior
5	–	Congestion management	–	–

decentralized mode, and have low distribution losses. Unifying all these components can make a microgrid. MG comes with an optimal solution by allowing multiple integration of RESs into the traditional power system. Based on their operation, they can be classified as grid-connected and autonomous mode MGs. The grid-connected MGs are normally controlled by grid parameters, but the islanded-MG operation is sophisticated due to intermittent nature of RESs and poses stability issues.[2] Despite their advantage of reducing growing carbon emissions, they have intermittent electrical outputs over distinct time zones (minutes, hours, and days) because of weather conditions, diurnal nature of sun, deviation in sun angles, etc. Also their integration leads to new technical challenges like weak inertia[3] problem in system followed by frequency excursions giving rise to frequency stability problems and hence imposes hindrances in making the balance between electricity generation and demand at all times which can eventually lead to blackouts in the system or some cascading problems. Consequently, imposing challenges in making balance between generation and load and requires proper communication system for smooth operation of MG.[4] In order to mitigate this challenge, there is a need of real-time data driven-based optimal planning and operation of such MGs which can improve reliability of electric supply because the previous research lacks in addressing this problem through this data-driven approach. Data-driven approach enables real-time operation of renewable rich power grids. Moreover, due to significant technological advancements along with encouraging government policies and significant environmental protection pressure, growing energy crisis, have shifted attention of people toward green energy like wind and solar energy. Appreciable amount of work has been done on renewable power generation, operation, and control but their stochastic nature has not been addressed and this problem has not been tackled using data-driven approach for their optimal operation which works on real-time basis and can give appreciable results when compared to traditional model-based methods.[4] So, to monitor the impact, renewable propelled power system is urgent need.

The system operators encounter numerous hindrances in maintaining and optimizing their grids as the electric grids are being stressed by the adoption of renewables and traditional model-based system control and operation do not give accurate results and require a proper communication network for their smooth operation. Nonlinearities in the operation of renewables create variations in load demand and uncertainties in power generation and consequently pose significant challenges for determining the optimal grid operational planning. Timely planning, analysis, and control of all the components in a power grid play a crucial role in achieving the resilient, sustainable, and secure smart energy infrastructure. Keeping these challenges in consideration, there is a need for assessing power system stability operation through data-driven methods which can open the possibility to incorporate system dynamics analysis and control mechanisms from measurements and realistic scenarios rather than model-based power planning stages.[5]

It is apparent that conventional methods of power system stability are prone to errors which are normally based on physical examination and mathematical modeling. These methods are also called model-driven methods. Being contradictory to efficiency and accuracy, these methods find least applications these days. Recently, data-driven methods like statistical analysis and artificial intelligence techniques present their appreciable results in grid stability assessment and control. For controlling, monitoring and coordination between smart equipment's in smart grid system, the huge amount of data acquired need to be analyzed, trained, and economically optimized bidirectional flow between utility and consumer. In order to develop economically optimal model for classification and prediction, the big data analysis with artificial intelligence-based machine learning (ML) techniques are essential and efficient.

Validations disseminate their outstanding results in computing, efficiency, accuracy, and decision-making process. Besides data-driven and model-driven techniques required for solving adequately the online stability assessment and control, we require some mathematical theory, data analysis also. Unifying both model-based and data-based techniques is considered as best and effective methodology. Data-driven methods result in high efficiency correlation analysis. Power systems data analysis is the new line of transformation, research, and identification, on line estimation of transient stability with renewables, PV power production and forecasting in MGs, load frequency control, and renewable integration impact assessment.[6] We intend to incorporate this methodology in existing researches of power system and MG architecture, this way we can develop smart grids. This unifying thought has also gained popularity in various other research areas of RDG like modeling, simulation, analysis, planning, abnormality detection, and energy management of electric power and MGs. To make increased penetration of RDG, we need proper planning for RDG which requires sufficient information for characterizing uncertainty of load and renewable output. Therefore, data-driven optimal planning is playing a key role in integrating RESs and is regarded as a modern technique in smart power system.[7] In this chapter, a novel dynamic data-driven technique is proposed for determining the real-time operation plan for renewable rich electric power MGs.

ML, as a subset of data-driven methods and computational intelligence, is currently an undergoing research and development, with the goal of extracting required knowledge from collected data sets and providing accurate forecasts and understanding using appropriate learning algorithms which can be used for power system operation and analysis, as well as bringing more flexibility and resilience to power generation systems. We can develop local DER controls for real-time operation of distributed networks utilizing ML approaches on optimal set points derived via optimization algorithms and recent developments in modeling and control of Mgs. The changes in the electrical structure are unprecedented, covering transformations in the energy landscape due to increase in active customers, new devices, and rapid technology, all of which have resulted in greater complexity and uncertainty, which has given rise to new issues. Better decision-making in the grid operation and planning paradigm is required to address this challenge. Furthermore, the modern grid is equipped with sensing and data collection facilities like smart sensors, advanced metering infrastructure.[8] This additional data, along with modern data analytics, appears to be a key component in enabling technology to make better decisions.

To improve electrical grid planning and operations, researchers are focusing their efforts these days on leveraging existing data and studying new data sources and data sets at a large scale. As a result, a data-driven strategy is regarded as a step toward the grid of the future. ML's purpose is to find a function that maps input data (explanatory variables) to observable Fig. 12.1 output (Response). ML techniques are frequently used in situations where traditional analytics are ineffective. ML is a broad category of algorithms that can be used to solve a variety of problems, including classification, regression, prediction, stochastic optimization, and so on.

The majority of ML algorithms are data-driven based, and current power systems include enough monitoring equipment to generate enough data to facilitate the implementation of these ML Algorithms. The data-driven stability assessment flowchart based on ML approach proposed for this study is shown in Fig. 12.1.

By just investigating the measurement data without too much prior information about the systems, ML algorithms ensure that model-free methods are discovered. Data management is relatively new topic for the power system. Due to availability of tremendous amount of data and strong

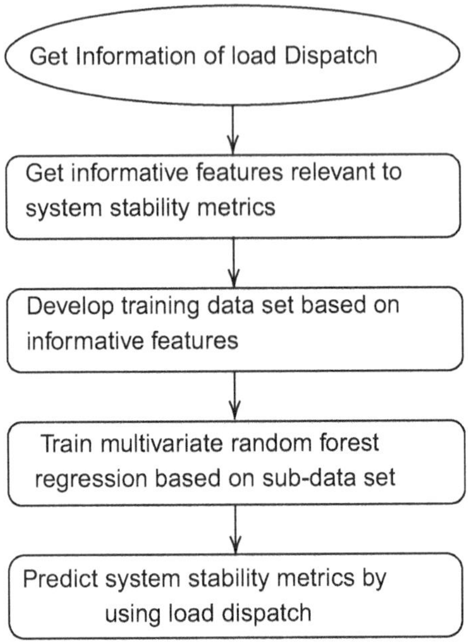

Figure 12.1 Flowchart based on data-driven approach.

computational power, ML algorithms find their appreciable applications which were not possible for power grid operators previously. Besides, ML models also aid system operators in making real-time decisions to reduce the effects of system occurrences.[9] There are various approaches that are semisupervised and selftaught. From technical perspective, ML techniques can be divided into four major categories: (1) supervised learning, (2) unsupervised learning, (3) semisupervised learning, and (4) reinforcement learning as shown in Fig. 12.2. In this chapter, we have presented the application of data-driven optimal approach to power system network rich in renewables and its appreciable results when compared to traditional-based

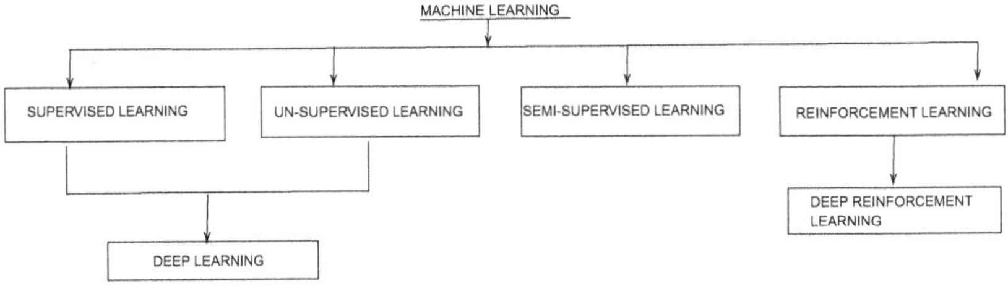

Figure 12.2 Illustration of machine learning framework.

techniques. The structure of this chapter is based on the sections, namely introduction, overview, potential applications of data-driven approach, MG structure and its control, networked MGs, intelligent energy storage devices, recent available techniques, operational aspects, and conclusion which are explained in a sequential fashion.

2. Overview of data-driven-based MG

The MG concept is regarded as a sophisticated method of integrating various DERs into the existing power distribution systems. MGs are being developed and deployed extensively from past few years due to their beneficial results compared to utility grid. However, researchers and developers are still struggling with their great efforts to improve distribution systems performance like minimizing losses, improving voltage profile, and thereby make the operation of modern power grids efficient. Recently, artificial intelligence-based ML techniques have received wide attention in modern grid studies due to their better results when compared to traditional power system control strategies, as the power sector is being pushed toward the digitization era to quickly adapt data-driven trends and large-scale renewable energy integration that is motivated by zero emissions.[10] For reliable, secure, and efficient operation of such resources leads to more renewable energies being integrated into the power system to replace conventional fossil fuels which helps to counteract global warming.

ML models also assist system operators to make real-time decisions to lessen the impacts of events hitting the system. Stability issues can be visualized properly using the proposed ML tool, specifically transient stability, frequency stability, and small signal stability. The proposed ML applications can effectively predict the stability margins for the above three mentioned stability metrics.[11]

3. Potential applications of data-driven analysis in smart power grids (Table 12.1)

PMUs and smart meters, generate a considerable amount of data due to improved communication and measuring technology which is not used in utility power grids for real-time operation of such grids. When used correctly, the acquired data can reflect the physical properties of operational power systems while also assisting in analysis and decision-making. ML-based algorithms are capable of extracting compact and meaningful information from high-dimensional data, and reinforcement learning (a branch of ML) can assist in decision-making by incorporating this retrieved data. This section provides a quick overview of ML applications in renewable rich power grids (Fig. 12.2).

4. Microgrid structure and its working

Owing to growing environmental concerns and inability of conventional systems to meet the present electricity demand, the thought of RE erupted and in the contemporary era, Fig. 12.3 the RE integration finds wide applications.[12] MG can be treated a mini-scale grid comprising of DGs, EES devices, and loads that are electrically bridged and hierarchically managed as shown in Fig. 12.3.

Depending on the system status or load requirement, a microgrid can operate in four operation modes: normal grid-connected operation mode, mode of transition from grid-connected operation to autonomous grid-connection, grid-reconnected mode.

4.1 Normal-grid tied operation mode

Grid-connected operation mode constitutes constant power, variable power, taking into account peak load shifting and optimal operation. During grid-connected operation mode, job of the controller is to ensure generation plans of each DG and make discharge plans of storage devices according to generation prediction and load prediction results of all the units.

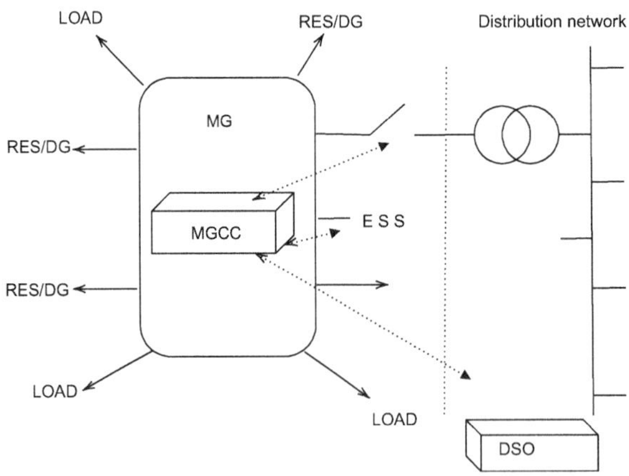

Figure 12.3 A microgrid example.

4.2 Transition from grid-connected operation to autonomous mode operation

Based on this operation mode, we have planned islanding and unplanned islanding mode. Planned islanding is sure to occur when there is command from control layer related to maintenance. While unplanned islanding may occur due to emergency conditions like severe faults hitting the system.

4.3 Autonomous mode operation

A standalone mode of MG consists of PV, wind, BSS, and DG as resources. A surplus amount of energy which the REs produce is used to charge the battery unit up to its rated capacity. Various parameters are taken into consideration here like optimal cost, optimal number of DERs required to meet the load demand. In this operation mode, the MG works independently and itself manages load generation balance. Mostly, the peak load is carried by this mode.

4.4 Grid-reconnected mode

Once the main power system has returned to healthy state, we can reconnect our MG with this main grid through PCC. RESs can be used in the MG architecture to provide power to remote locations when expansion of the transmission infrastructure is not cost-effective. MG enables the development of a viable alternative to the existing centralized power grid. The MGs are miniature electrical distribution systems that can link numerous DGs, consumers, and energy storage systems together. Before the development of MGs, it is necessary to model MG with distributed sources and storage devices. To ensure seamless integration of DERs is a prerequisite before the formation of MGs. The MGs frequency and voltage are kept at nominal values to ensure the system's stability.[13]

Due to poor inertia problems in an MG, the primary level cannot maintain smooth nature of frequency and voltage of the system when MG is disturbed by either load changes or connection/disconnection of units. While running in a steady state, the secondary control is used to reduce frequency and voltage irregularities and maintain the balanced operation of MGs. Energy storage systems are required to eliminate the intermittent character of RESs. The energy storage system, which balances the power flow at the start of the autonomous mode of operation, is a critical component of MG. It also aids in the improvement of power quality and voltage profile, as well as the management of power flow to and from the device.[14]

4.5 Layered control strategies

The MG has a multilayered gradable edifice and needs different control strategies critical for the safe operation of MG in both grid-connected and autonomous modes, as well as successful disconnection/reconnection operations. These controllers must ensure that these operations run smoothly and that the system is able to operate in the desired modes.[13] The MG can be controlled by control loops classed as community, auxiliary, extremis, and global controls, similar to large-scale classical power systems. The secondary, global, and emergency controls, in contrast to community control, which operates without communication, require a communication bus as shown in Fig. 12.4. Decentralized controllers are used for local controls, while centralized controllers are used for secondary and emergency controls. In traditional functioning, the MG is connected to the primary distribution grid, either partially supplying or injecting power into it. The primary grid and local DGs can provide power to the loads in a grid-connected manner, allowing for a steady supply to customers.[15] Assume, however, that any anomaly in the conventional grid arises. As the entire network is (Fig. 12.4) operated in an organized form of an MG through an MGCC, an autonomous operation mode can be enabled.

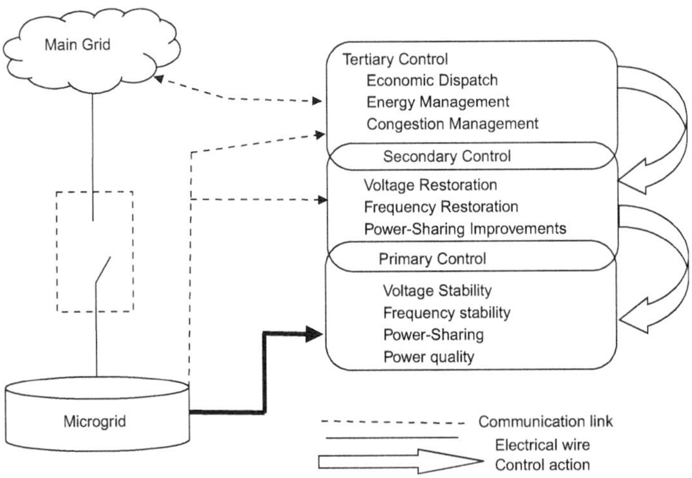

Figure 12.4 Hierarchical control of microgrid.

5. Networked microgrids

The establishment of networked MG concept is aided by recent advancements in MG technology. In networked MGs, many MGs are connected to each other. Through networked MGs, the reliability and resilient benefits can be achieved through energy exchange mechanism. Moreover, NMGs facilitate ancillary services. However, several stability issues arise while interconnecting several MGs which need to be addressed in order to ensure smooth operation of power system. The data-driven approach for same system can facilitate real-time monitoring, control, and operation of renewable penetrated power system using different learning algorithms.

6. Intelligent energy storage devices

Maintaining a balance between generation and loads in an MG is a difficult task. We know that integrated RESs are intermittent, and we cannot forecast when they will be at their best. We are supposed to have efficient storage which can regulate our frequency deviations. As a result, intelligent storage devices are seen as the smart solution for addressing load and generation mismatches and facilitating the adoption of renewable energy sources by storing excess energy while demand is low and harvesting it when peak demand rises. Energy storage systems build virtual inertia in weak MGs as the MGs lack the inertia thereby can stabilize voltage and frequency excursions. Different energy storage devices are available like flywheel, SMES, batteries, and hybrid storage devices. The optimal usage of these storage devices depends on duration of disturbance hitting the system for small and persistent disturbances we can use batteries but on the contrary for large disturbance of small duration we can use SMES, fly wheel. The optimal operation of storage devices like efficiency and life period is dependent on their physical pattern like state of charge and health. Apart from storage devices, conservative voltage reduction also helps in smooth operation of weak grids by shifting peak loads.

7. Operational aspects for distribution system

Data-driven approaches in power systems have opened the option of incorporating system dynamics analysis and control mechanisms from measurements, as well as realistic scenarios, rather than model-based power planning stages, as a result of the increasing complexity of operations. The present frontier of innovation, investigation, and study in the field of power systems data analysis is on the rise. This type of data analysis has been used to address a variety of system dynamics, including online prediction of transient stability with renewables, forecasting photovoltaic power output in MGs, and assessing the impact of renewable integration.

8. Recent available technologies and their outstanding results

The energy system is in the midst of a rapid transformation to a more sustainable future. DERs will be critical in transforming the current power grid into the smart grid of the future. They can assist in the integration of dispersed renewable power generation and storage amplitude into grid, improving grid resilience while reducing reliance on carbon-emitting fossil fuels for power generation. Because of the fluctuation of power generation from renewable energy sources, as well as the lack of predictability, a focus has been placed on developing data-driven strategies to integrate those renewables. Voltage and frequency unpredictability are two additional issues for utility providers as a result of these resources within the system.[13,15–19] The primary objective of data-driven methods in power system is to develop methodology that can facilitate smart use of data for power system operations. The proposed research focuses on recognizing real-time power system dynamics from data in order to improve real-time monitoring. When compared to previous mathematics innovative model-based methodologies, the proposed approach is computationally efficient. ML, as a subset of data-driven methods and computational intelligence, is ongoing research these days has been in continuous allowing for the extraction of required information from collected data sets and the provision of accurate predictions and insights that can be used for power analysis and can increase the flexibility and resilience of power generation. By using ML techniques on optimal set points obtained from optimization, we can design local DER controls for real-time operation of distributed network as shown in block diagram of Fig. 1.2. The changes in electricity industry are unprecedented, encompassing shifts in energy scenario, rising active consumers, new devices, and fast technology, As a result, complexity grew, and uncertainty arose, posing new obstacles. Better decision-making in the grid operating and planning paradigm is required to address this challenge. In addition, the grid will be equipped with sensing and data collection capabilities. This new data, along with fresh data analytics, emerge as a critical aspect in enabling technology to make better decisions.

8.1 Data-driven-based forecasting and fault detection

Features concealed in the acquired data can be retrieved using a data mining approach. These features are useful for power system analysis because they represent the relationships between the inputs and outputs. Renewable energy sources like solar, wind being intermittent in nature, need accurate forecasting facilities so that we have least uncertainty in generation and power fluctuations. The renewable energy sources are highly dependent on forecasting model, the proposed forecasting model can be developed by two methods, namely, feature extraction and regression.[16] The feature is firstly extracted through a memory network and sent to learning algorithm for forecasting analysis. Apart from generation forecasting, we need load prediction also which is crucial for energy management of power systems. Residential loads are supposed to be of high uncertainty resulting from human activities when compared to industrial and commercial loads. Data-driven-based methods enable efficient isolation and rapid system recovery benefit when power system is subjected to contingencies.

8.2 Energy management

The primary goal of power systems is to maintain a cost-effective, stable, and long-term energy supply. To complete this work, you will need to keep track of and handle unpredictable aspects, coordinate multiple resources, and create effective algorithms. Several technical challenges arise when high penetration of renewable energy occurs, as frequent fluctuation and poor predictability of DERs pose challenges in their stable operation. The ML approach come with better results to manage the uncertainty from RE. EMSs perform an essential role in coordinating the diverse DERs in order to meet various operating objectives. Due to application of ML approach, we can extract the useful information from collected data and hence can provide fast and accurate results of uncertain renewable generation and load consumption. The ML-based energy management strategy (EMS) is supposed to give promising results when compared to traditional model-based systems. As a result, there is an increasing need to build an EMS that al-lows surplus energy to be transferred to the grid or MGs via a fuzzy-based peer-to-peer energy exchange algorithm with dynamic pricing and MGs future information.

8.3 Optimal power system control

Normally, the control problems in a power system consist of tracking and controlling problems. The objective of a desired problem is to chase a reference signal and objective of later is to satisfy a specific control function. Data-driven-based approach is best solution for solving such problems, exclusively in areas where system information is unknown and only data are available. Undoubtedly, increasing utilization of RE, for example, wind and solar photo voltaic makes the environment eco-friendly and reduce the operational costs. But, at the same time, their integration creates the challenging issues, one being how to maximize the power output of renewable generators under the stochastic environmental conditions. This problem we call as maximum power point tracking (MPPT). The MPPT-based algorithms work well when we have significant knowledge of systems like wind speed, solar radiance information. Currently, we have sufficient data collecting devices. Therefore, by incorporating ML-based learning algorithms, we can develop the predictive model of the system from which we can predict the stability of the proposed system. Moreover, the proposed study is able to overcome the hindrances found in RE integration and traditional methods and can work better in improving the transient performances.

9. Conclusion and future scope

Wide area development of communication technology and data science provides the means for ML applications in power grids. This study proposes the data-driven-based ML applications to power distribution systems as already discussed in introduction (Section 1) of this chapter followed by overview of MG (Section 2) wherein components of MG are highlighted. Section third of this chapter describes applications of data-driven methods to MG, like forecasting, planning, monitoring, and energy management. In section fourth, MG structure and its working is described and its various required control strategies. In section fifth, networked MG structure is mentioned and in the last section of this chapter, the idea of recent available technologies is apparent. The ML-based methods for intermittent renewable-based power grids are providing promising results for stability requirements. In general, the idea of using ML technique is to exploit the available data using suitable algorithms. These ML-based algorithms find their applications in modern power systems, which are having uncertainty, high data through advanced metering infrastructure, ML techniques give excellent results, and here we do not need explicit system models are not needed and control stability can be ensured. To avoid repeating manual and traditional work in power grids, auto Ml is a fascinating solution, which aims at automatically implementing the process of applying ML through an ML model. In future work, we will extend this technique to distribution side of power system for better operation and control of system network and also to plug-in hybrid electric vehicle.

References

1. Kumar, Barjeev. An optimal multivariable constrained nonlinear (MVCNL) stochastic microgrid planning and operation problem with renewable penetration. *J IEEE Syst J*. 2020;14:4143–4154.
2. Bagheri A, Wang J, Zhao C. Data-driven stochastic transmission expansion planning. *IEEE Trans Power Syst*. 2017;32(5):3461–3470. https://doi.org/10.1109/TPWRS.2016.2635098.
3. Ratnam KS, Palanisamy K, Yang G. Future low inertia power systems: requirements, issues and solutions - a review. *Renew Sustain Rev*. 2020;124.
4. Sahoo SK, Sinha AK, Kishore NK. Control techniques in AC, DC, and hybrid AC-DC microgrid: a review. *IEEE Trans Emerg Sel*. 2018;6(2):738–759. https://doi.org/10.1109/JESTPE.2017.2786588.
5. Sun M, Cremer J, Strbac G. A novel data-driven scenario generation framework for transmission expansion planning with high renewable energy penetration. *Appl Energy*. 2018;228:546–555. https://doi.org/10.1016/j.apenergy.2018.06.095.
6. Tan J, Zhang Y. Coordinated control strategy of a battery energy storage system to support a wind power plant providing multi-timescale frequency ancillary services. *IEEE Trans Sustain Energy*. 2017;8(3):1140–1153. https://doi.org/10.1109/TSTE.2017.2663334.
7. Yan Z, Xu Y. Data-driven load frequency control for stochastic power systems: a deep reinforcement learning method with continuous action search. *IEEE Trans Power Syst*. 2019;34(2):1653–1656. https://doi.org/10.1109/TPWRS.2018.2881359.
8. Zhang Y, Xu Y, Dong ZY. Robust ensemble data analytics for incomplete PMU measurements-based power system stability assessment. *IEEE Trans Power Syst*. 2018: 1124–1126. https://doi.org/10.1109/TPWRS.2017.2698239.
9. Zheng L, Hu W, Zhou Y, et al. Deep belief network based nonlinear representation learning for transient stability assessment. In: *IEEE Power and Energy Society General Meeting*. 2018:1–5. https://doi.org/10.1109/PESGM.2017.8274126.
10. Zhu L, Lu C, Dong ZY, Hong C. Imbalance learning machine-based power system short-term voltage stability assessment. *IEEE Trans Ind Inf*. 2017;13(5):2533–2543. https://doi.org/10.1109/TII.2017.2696534.
11. Chen Z, Xiao X, Li C, Zhang Y, Hu Q. Real-time transient stability status prediction using cost-sensitive extreme learning machine. *Neural Comput Appl*. 2016;27(2): 321–331. https://doi.org/10.1007/s00521-015-1909-9.
12. Du E, Zhang N, Hodge BM, et al. The role of concentrating solar power toward high renewable energy penetrated power systems. *IEEE Trans Power Syst*. 2018;33(6): 6630–6641. https://doi.org/10.1109/TPWRS.2018.2834461.
13. Katiraei F, Iravani MR. Power management strategies for a microgrid with multiple distributed generation units. *IEEE Trans Power Syst*. 2006;21(4):1821–1831. https://doi.org/10.1109/TPWRS.2006.879260.

14. Gómez-Expósito A, Conejo AJ, Cañizares C. *Electric Energy Systems: Analysis and Operation*. Spain: CRC Press; 2016:1–648. https://doi.org/10.1201/9781420007275.
15. Impact of multi- microgrid communication systems in islanded operation. *Proc IEEE*. 2011:1–6.
16. Duchense L, Karangelos E, Wehenkel L. Recent Developments in machine learning for energy systems reliability management. *Proc IEEE*. 2020.
17. Gurusinghe DR, Rajapakse AD. Post disturbance transient stability status prediction using synchro phasor measurements. *IEEE Trans Power Syst*. 2018;31:3654–3664.
18. Pan S, Morris T, Adhikari U. Developing a hybrid intrusion detection system using data mining for power systems. *IEEE Trans Smart Grid*. 2015;6(6):3104–3113. https://doi.org/10.1109/TSG.2015.2409775.
19. Vijay AS, Doolla S, Chandorkar MC. Unbalance mitigation strategies in microgrids. *IET Power Electron*. 2020;13(9):1687–1710. https://doi.org/10.1049/iet-pel.2019.1080.

CHAPTER THIRTEEN

Short-term load forecasting in the presence of grid uncertainties using new methods based on deep learning

Seyed Mohammad Shobeiry, Sasan Azad and Mohammad Taghi Ameli
Faculty of Electrical Engineering, Shahid Beheshti University, Tehran, Iran

Contents

Smart Electrical and Mechanical Systems
ISBN: 978-0-323-90789-7
https://doi.org/10.1016/B978-0-323-90789-7.00006-3

1. Introduction

Over the past decade, the electricity industry has made extensive investments in smart grids (SGs). In an SG, information and energy are exchanged in a two-way transmission. The SG needs an uninterrupted power supply to the load, which requires a suitable tool to forecast the load with minimal error. Load Forecasting (LF) is utilized to control network operation and planning determinations.[1] In addition, LF helps stabilize an organized SG and minimizes the electricity generation cost. Different variables affect the data and complicate the forecast. Thus, the load at a given hour is affected by yesterday's load data, weather, demographic data, customer type, and econometric data.[2]

Therefore, the consumption load, especially residential load, is uncertain due to its dependence on various factors such as weather. On the other hand, issues related to smart homes and demand-side management have added to the uncertainty of this variable and made its prediction more complex. Due to the need to solve complex relationships in SG problems, many applications have been provided for methods based on artificial intelligence (AI), such as machine learning (ML) and deep learning (DL). For example, in reference[3], a combined forecasting approach called quantile regression averaging (QRA) is presented for the probabilistic forecasting of ambient temperature. This process provides complete information about future uncertainties. The proposed method is more appropriate than the quantile k-nearest neighbors, quantile egression forests, and basic quantile regression inferred from the quantile score, Winkler score, and reliability plots. Also, the QRA has been successful in LF and electricity spot prices. Reference[4] applies the QRA to accomplish a probabilistic forecast of PV generation using its historical record. The proposed model is used in the short-term forecasting of PV generation for the four major seasons up to 2 weeks ahead. ML models are basic AI algorithms that extricate patterns to form rational determinations from raw data.[5] DL is known as a subfield of ML that has a set of intelligent computational algorithms. Deep neural networks (DNN) provide a comprehensive solution for modeling complex nonlinear relationships between input and output through their many layers. Therefore, using DL methods in LF can be a new and efficient solution.

Issues discussed in the field of ML are divided into supervised and unsupervised problems. Supervised issues have labeled datasets, and unsupervised issues have unlabeled datasets. However, an unsupervised learning algorithm is used for a data set with many features to learn the valuable properties of the data set structure. Data sets are always divided into training and test data in learning algorithms. The ML model extracts specific patterns from the raw data or, in other words, is trained from the raw data with the help of training data sets and evaluates its output validity with the help of test data sets. Improving the speed and accuracy of LF has become an essential milestone in researches. Factors that influence the execution of LF are due to different viewpoints. From a modeling point of view, the correct choice of models and parameters significantly impacts LF performance. Therefore, it is necessary to facilitate the LF performance, especially in parameters and models. Therefore, this chapter, in comparison with the literature, while introducing different aspects of LF, focuses on one of the critical and practical components of LF, namely forecasting models, to raise the DL models used in this field in the presence of existing uncertainties. After introducing the details of DL-based models for short-term load forecasting (STLF), a comparison is made according to the forecasting accuracy criterion to create a clear vision for the models in future research. The sections presented in this chapter are as follows: In the introduction section, the context, importance, and the main contribution of this chapter compared to literature were described. The following section will introduce various aspects of LF, such as benefits, influencing factors, challenges, and existing uncertainties. Finally, in the third section, explanations of the evaluation criteria and classification of LF types will be provided. The fourth section will discuss DL methods, performance details, and their comparison for the LF application.

2. Different aspects of LF

2.1 LF benefits

LF has significant advantages that have attracted many researchers in this field. From the earliest electricity generation, service organizations have had an essential question to circumscribe the demand for the following periods to adjust extending demand and confined resources. Although renewables have alleviated the problem of resource management, energy gathering methods are still costly and problematic. If demand is estimated in advance, the dispatch of generation units can also be optimized; hence, the term "LF" emerged. There are additional advantages to LF, which include the following:

(1) LF helps plan resources such as fuels needed in the generation sector, continuous electricity distribution, and affirming economic generation.

(2) Understanding future demand will help utilities make cost-effective economic decisions and planning.
(3) LF ensures maximum optimal power plants use, which helps reduce carbon emissions and reduce fossil fuel consumption.
(4) LF helps make decisions and plans about maintaining power systems.
(5) LF assists plan the construction, type, capacity, and location of power plants by providing estimates of transmission and distribution infrastructure costs.

2.2 Factors affecting LF

During the training process, the LF model must learn the behavior of the electric load pattern according to the collected load data, which is the most crucial component of the model. Preprocessing on load data must then be performed to achieve the training objectives. In the preprocessing step, nonexistent data and out-of-range data are created and edited. Adequate input data in the LF model include electrical load data, historical weather data, historical event data, and so on. The collected data are analyzed to select and extract the most influential variables in the problem. Then, according to the nature and dimensions of the problem, several models are nominated for LF. Among these models, the best model with the appropriate performance for execution is selected. Most of the factors affecting LF accuracy are based on location and equipment, which need to be considered when designing an accurate LF model.

As mentioned, the experimental forecasting process relies on several factors that affect the accuracy of the forecast, and these factors must be carefully selected. Some of the most critical factors include weather, time, and economic factors. Historical load data, weather data, and events are collected through various technologies and algorithms used in sensors, smart meters, and data servers for analysis and model entry.[6] Of course, it is necessary to mention that the choice of model also requires a lot of attention because different algorithms based on these factors use other parameters. Essential and effective factors on data collection and model preference include the following:

2.2.1 Time

Time is the most crucial factor in an LF problem that can be considered in different sequences. Ruzic et al.[7] from the analysis of the load curves in a sample network, concluded that the load curves have the characteristics of "time of day" "day of the week" "week of the month" and "month of the season" Also, the current data and the data of the past days for a specific place help forecast accurately. In addition, the time frame for LF is essential because it determines the amount of data needed to implement the process.

2.2.2 Weather

It can be said that the weather situation is one of the most independent variables in the LF problem and has a more significant impact on urban and agricultural consumers. For example, with changes in the weather and the use of heating and cooling devices, electricity consumption increases. Also, a sudden drop in temperature can reduce consumption and thus, LF becomes excessive. Therefore, weather forecast results are used to forecast future load demand by different models. Some of the most critical weather factors influencing the LF problem include humidity, temperature, and dew point temperature. Service companies also widely use the Temperature-humidity Index and Wind Chill Index.

2.2.3 Economic

Demand-side management, electricity prices, technological advances, and the rate of industrialization are the critical economic factors that significantly impact the system's average load and peak load.[8] In addition, power consumption behavior, change in electricity tariffs, type of equipment, area population, age of equipment have an influential role in determining the forecast accuracy. These factors need to be analyzed and calculated to forecast long-term loads. People's economic behavior is deemed to extrapolate the demand for electric load.[9]

2.3 Challenges and uncertainties of LF

The previous section discussed the factors affecting LF as input variables for developing an LF model. Sometimes it isn't easy to consider all the factors due to the unavailability of data. Different area data are collected from relevant sources to elaborate an LF model in a specific area. For example, weather information is collected from the local meteorological office. Data aggregation is also a challenge due to the unavailability of appropriate data, the existence of data anomalies, and the loss of data values (more than 5%). The presence of one of these factors can reduce forecasting accuracy. Hence, possible solutions to address such challenges have been proposed by researchers. Accurate LF can be a great potential for cost savings for utility companies, while unintended errors can lead to significant financial and infrastructure losses. Concurring to Haida et al.[10] forecasting errors can increase the cost of electricity generation. Only by reducing the average absolute percentage error (mean average percentage error [MAPE]) by 1% and accurately LF, by reducing the generation cost by about 0.1%–0.3%, an efficiency improvement of 3%–5% on the generation side is achieved.[11] In recent years, state-owned electricity companies have shown great interest in generating electricity based on renewable energy. The growing number of renewable energies poses significant challenges. The probabilistic and random behavior of power generated from wind and solar sources makes it difficult for management and supply companies to adopt this generation type to the load. Research has been done to develop an innovative power dispatch method that appends intermittent sources such as solar, wind, etc.[12–15]

For years, researchers' primary goal has been to develop a high-precision LF model to improve the efficiency of the power grid and the revenue of power generation and service companies. In recent years, many improved methods have been proposed, but still, many challenges prevent the proper conduct of this research and achieving the appropriate accuracy of the models. Some of these challenges are listed below:

(1) Consumers in different regions use different types of meters with additional tariffs. In addition, the behavior of using electricity varies between customers in different situations. The forecasting tool must clearly understand the use of the systems and create a separate forecasting model for each measurement system. Finally, their aggregation will make the final forecasting value.

(2) One of the most exciting and influential factors in LF is the weather condition. Weather conditions are not always predictable, and a significant error in LF may occur in the abrupt weather change. Also, an extensive energy system has different regions that have different geographical and weather conditions.

(3) Accurate LF is a complex issue because of the combination of several factors that affect electricity demand.

(4) Electricity prices, or in other words, consumer costs, are other challenging factors in LF. A sudden change in pricing and the interconnected nature of demand based on price change leads to a shift in consumer behavior.

(5) The occurrence of sudden disturbances such as unexpected and transient errors in the power system affects the load. These events generate specific data that need to be considered separately for model design. They should not be evenly distributed with standard data.

(6) Accurate demand forecasting based on temperature change, humidity, and other factors affecting electricity consumption is nonlinear and complex.
(7) The electricity distribution company may suffer technical and financial losses if it encounters a high STLF error.

Various challenges caused designers to focus on the multiple components of an LF problem, such as model, input data, algorithm parameters, and selecting the appropriate model for LF.

3. LF classification and evaluation criteria

In terms of periods and practical intervals, LF can generally be classified into four categories, which include

a. very short-term load forecasting (VSTLF),
b. short-term load forecasting (STLF),
c. medium-term load forecasting (MTLF), and
d. long-term load forecasting (LTLF).

VSTLF is used to forecast load from a few seconds to a few minutes.[16] The recently observed load pattern is utilized rather than modeling the relations between time, weather conditions, load, and other factors. Also, in this period, information about the economy, temperature, and type of use can be optional. VSTLF methods include artificial neural network models, genetic algorithms, and moving averages.

STLF plays a vital role in the system's operation, so it is used for the forecast from a few minutes to a few hours. This forecast is the primary source of information for all activities related to the daily and weekly process of generation sector planning.[17] Research shows that interest in developing short-term forecasting models has recently increased. A good understanding of the factors affecting the load, such as weather conditions, season, type, and time of day in a particular region, will lead to more appropriate implementation of the short-term forecast model. Due to the dependence of factors and their effect on the electric load, the combination of factors and electric load is the main issue that must be considered. MTLF is commonly used to forecast loads over days to months.[18] LTLF is also used for several weeks to several years.[19] The problem of LTLF can have wide dimensions; some of the factors considered in this type of forecast include economic factors, historical load, weather data, number of customers in different categories, features of Home Appliances. Table 13.1 provides a comparison of LF methods based on period and employment.

Table 13.1 LF methods classification.

Methods	Period	Employment
VSTLF	Few minutes	Generation forecasting & distributions planning
STLF	Few hours	Maintenance planning & spinning reserve allocation
MTLF	Few days to months	Seasonal LF
LTLF	Few years	Generation growth planning

STLF is the foremost broadly utilized and well-known strategy due to its inherent connection with other types of forecasts and playing a pivotal role in formulating economic and safe procedures for the power system. It is also possible to convert the forecast intervals by adding or decreasing different input variables. For example, by adding loads of the previous hours to the VSTLF model, it can be converted to the STLF model or converted to MTLF or LTLF by adding economic variables to STLF.

Many criteria are used to evaluate the accuracy of different LF models. All of these criteria represent the difference between the predicted value and the actual value. Table 13.2 presents the most widely used criteria in recent research.

In Table 13.2, y_i is the actual value of the electric load, y_i' is the forecasted value of the model, and n is the number of samples. Each of these criteria has its focal points and impediments. RMSE accentuates more significant errors than minor errors and gives a quadratic loss function. MAE is used to measure the average error. MAPE is recommended to avoid dependence on the scale. Small and near-zero denominators are considered one of the weaknesses of MAPE, but this is not important for LF problems because the load at the aggregate level is rarely close to zero.[20] Finally, it should be noted that it is complicated to compare the results with different criteria, and the value of these criteria varies according to different data sets and models.

Table 13.2 The most widely used evaluation criteria.

Criteria	Formula
Mean absolute error (MAE)	$\frac{\sum_{i=1}^{n} \lvert y_i' - y_i \rvert}{n}$
Root Mean Square Error (RMSE)	$\sqrt{\frac{1}{n} \sum_{i=1}^{n} \left(y_i' - y_i\right)^2}$
Mean average percentage error (MAPE)	$\frac{1}{n} \sum_{1}^{n} \frac{(y_i - y_i')}{y_i}$

4. DL methods

The term DL was first coined by Dexter[21] in 1986. After the gradual improvement of the algorithm for several years, this concept is now known as a DNN. The structure of DL is inspired by the functioning of the human brain.[22] The typical neural network has a shallow and superficial structure consisting of input, hidden, and output layers. In contrast, a DL architecture has more hidden layers than a multilayer perceptron network (MLP). Due to the ability to model nonlinear relationships and the high learning power of the model, DL is widely used in prediction programs.[23] DL is effective in solving complex functions with high numerical dimensions and highly variable functions. One of the main drawbacks of this model is the problem of overfitting due to the need for a large number of layers to achieve accurate output. Of course, it should also be noted that DL often requires a lot of execution time due to the algorithm's complexity.

Ryu et al. have presented two DNN designs to learn the complex relationships between variables of weather, history, and previous consumption.[23] In reference[24], a deep belief network-based model (DBN) is proposed to forecast the hourly load. By increasing the volume and variety of data, Shi et al. could solve the overfitting problem in the DL algorithm. A new pooling deep recurrent neural network (PDRNN) is introduced that creates a set of categorized inputs from load profiles. The projected strategy comprises two fundamental steps: (1) load profile aggregation and (2) implementation of STLF with a deep recurrent neural network (D-RNN). This method has the necessary efficiency in learning the existing uncertainties such as load variety and weather conditions. The suggested method performs better than the classic ARIMA, support vector regression (SVR), and D-RNN procedures according to the RMSE.[25] A framework based on customer behavior and consumption sequence retention using long short-term memory (LSTM) algorithm to address the uncertainty is presented.[26] RNN and its various subbranches are widely utilized for STLF in the residential sector.[27]

4.1 Autoencoder

The autoencoder is utilized to copy the input neurons in the output neurons by transferring neurons from hidden layers (h).[5] The main components of an autoencoder network architecture are based on the function of the encoder $h = f(x)$ and a decoder for reconstruction $\hat{x} = g(h)$. Thus, the reconstructed output is $\hat{x} = g(f(x))$, which copies the input data. The autoencoder equation presented in Eq. (13.1):

$$\hat{x} = g(Wx + b) \tag{13.1}$$

g is the activation function, W is the weights, x is the input, and b is the bias. This algorithm is commonly utilized to reduce the size, feature learning, or reconstruct corrupted data[5]; the autoencoder architecture is presented in Fig. 13.1.

4.2 Recurrent neural network (RNN)

The recurrent neural network (RNN) has an internal memory that changes the neuron state according to the prior input. In other words, the recurrent neural network is also called the sequential data processor. The recurrent neural network is usually trained by backpropagation. However, it is typically weak due to the disappearance of the descending gradient.[28] The deep RNN creates more layers in the recurrent neural network architecture, which offers significant benefits.[29] The recurrent neural network computational graph has N consecutive inputs x^i and output y^i consecutively when the sequence time from t to τ is specified.[25] In the following mathematical representation (relationships 2 to 4) of the computational graph of the recurrent neural network model, the number of layers is specified as l from 1 to

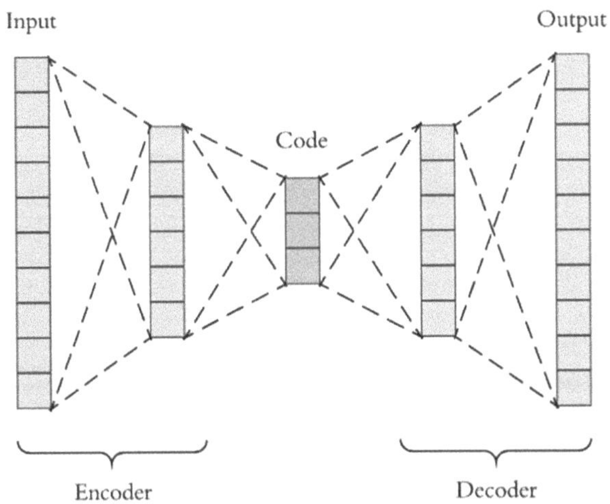

Figure 13.1 Autoencoder algorithm architecture.

N. The activation function for l is shown as h_l also $x(t)$ is input and $y(t)$ is output. Also, W and b, respectively, indicate the weights and biases.

$$a_1(t) = b_1 + W_1 \times h_1^{(t-1)} + U_1 \times x^{(t)} \quad (13.2)$$

$$a_l(t) = b_l + W_l \times h_l^{(t-1)} + U_l \times h_l^{(t)} \quad (13.3)$$

$$y(t) = b_N + W_N \times h_N^{(t-1)} + U_N \times h_N^{(t)} \quad (13.4)$$

4.3 Long short-term memory (LSTM)

As referenced previously, the recurrent neural network fails to address the disappearance of the descending gradient. LSTM has a long-term memory that can solve this problem in recurrent neural networks. The LSTM has inner loops that are used to store information. There are five critical elements in the LSTM computational graph: (1) input gate, (2) forgetting gate, (3) output gate, (4) cell, and (5) state output.[28] Fig. 13.2 shows the architectural diagram of an LSTM cell. A computational representation of an LSTM cell is given in Eqs. (13.5)–(13.10).

$$x_i = \sigma(x_t W_{x_i n} + h_{(t-1)} W_{x_i m} + b_i) \quad (13.5)$$

$$x_f = \sigma(x_t W_{x_f n} + h_{(t-1)} W_{x_f m} + b_f) \quad (13.6)$$

$$x_o = \sigma(x_t W_{x_o n} + h_{(t-1)} W_{x_o m} + b_o) \quad (13.7)$$

$$U = \tanh(x_t W_{Un} + h_{(t-1)} W_{Um} + b_U) \quad (13.8)$$

$$C_t = x_f \times C_{t-1} + x_i \times U \quad (13.9)$$

$$h_t = x_o \times \tanh(U) \quad (13.10)$$

where x_i is the input of input gate, x_f is the input of forgetting gate, x_o is the input of the output gate, U is the update signal, C_t is the state value at time t, and h_t is the LSTM cell output. The memory state can be modified by deciding the input gate using the sigmoid function with on/off mode. If the input gate value is minimal and close to zero, no change is made to the mode cell memory C_t. The stacked LSTM in the network model can be displayed as multiple LSTM layers.

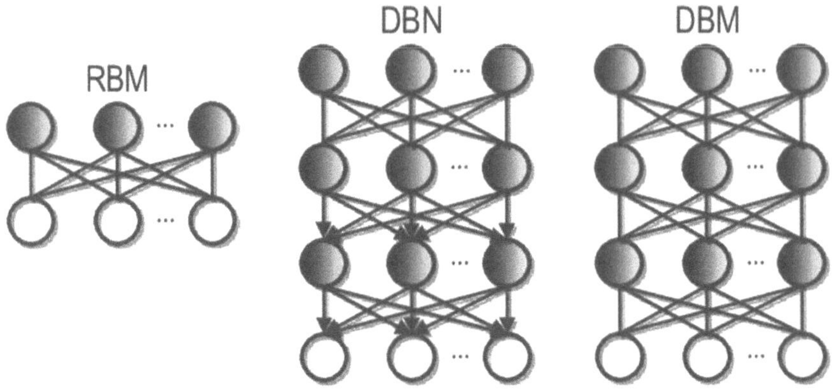

Figure 13.2 LSTM architectural diagram.

4.4 Convolutional neural network (CNN)

CNN can be thought of as part of a feeder neural network that mimics the structure of human nerve cells. CNN is commonly used to process data dimensions.[5] This type of DL is widely used in speech, auditory, and visual processing. Time series data has a one-dimensional structure in a time interval, and image data has a two-dimensional structure of pixels.[5] The mathematical function of convolution is used in at least one layer of CNN.[5] The mathematical relation of convolution is presented in Eq. (13.11):

$$s(t) = (x^*w)(t) = \int x(a)w(t-a)da \tag{13.11}$$

Which w is called kernel filter, x is called input, and s is output, which is called feature mapping for continuous-time t. The function of discrete-time convolution is as follows Eq. (13.12):

$$s(t) = (x^*w)(t) = \sum_{a=-\infty}^{\infty} x(a)w(t-a)da \tag{13.12}$$

For a two-dimensional image I with dimensions i and j, and a two-dimensional kernel filter K with dimensions m and n, convolution as follows Eq. (13.13) is called cross-correlation[1]:

$$s(i.j) = (I^*K)(i.j) = \sum_m \sum_n I(i+m.j+n)K(m.n) \tag{13.13}$$

CNN is made up of three primary layers; the first layer has parallel convolution layers. The second layer is the detection stage with a rectified linear activator. The third stage is the pooling function layer.[5]

4.5 Restricted Boltzmann machine (RBM)

RBM model is one of the generally famous deep probability directionless Algorithms.[5] RBM has two primary layers, the first containing visual inputs and the second containing hidden variables. RBM is ordinarily stacked to create deeper layers by building one portion on the beat of the other. Fig. 13.3 appears an RBM for a probability two-layer directionless model.

4.6 Deep belief network (DBN)

DBN has multiple hidden layers based on the accumulation of RBMs and is taught using the backpropagation algorithm.[23] In this architecture, there are connections between each unit in consecutive layers, but the units of each layer have no internal connection.[5] Fig. 13.3 shows a three-layer structure consisting of two hidden layers and one visible layer for DBN. In this structure, the top two layers are directionless; the connection between the other layers must be directed to the visible layer.[5]

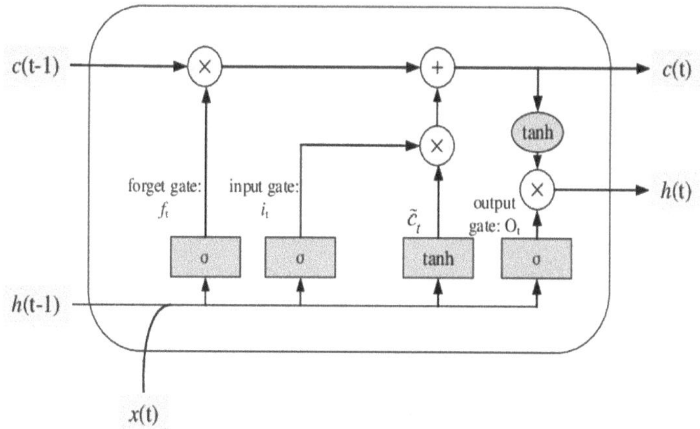

Figure 13.3 Various architectures include RBM with directionless connections between visual inputs and hidden variables. DBN with directional connections to observable inputs and directionless connections for hidden layers and DBM with directionless connections for one visible input layer and several hidden layers.

4.7 Deep Boltzmann machine (DBM)

DBM architecture has several hidden layers with complete directionless communication and a visible layer.[5] The main difference between DBM neural network and RBM is the more significant number of hidden layers in DBM. Fig. 13.3 shows the DBM neural network architecture.

5. Forecasting results analysis and compare

In many articles, the Root Mean Square Error (RMSE) is used to evaluate the proposed method performance compared to other existing methods. RMSE is commonly used to determine prediction accuracy error. The criterion for measuring the accuracy of the DL methods is presented in Eq. (13.14):

$$\mathrm{RMSE} = \sqrt{\frac{1}{N}\sum_{n=1}^{N}(y - x)^2} \tag{13.14}$$

N is the number of time series instances, y is the forecasted output time series and, x is the input time series. In Table 13.3, most of the reviewed articles are summarized in comparing the proffered method with the standard benchmark method. Table 13.3 also shows the RMSE reduction percentage compared to other methods. In Table 13.3, the decrease of RMSE in the DBN group method and SVR compared to the fuzzy neural network (FNN) is significant. In addition, the PDRNN method shows an excellent reduction compared to ARIMA. Also, CNN with the k-means has shown a very high RMSE reductions percentage.

Table 13.3 Percentage reduction of RMSE compared to the proposed approach and the standard benchmark approach.

Common benchmarking approach	Proposed approach	RMSE decrease percentage
ARIMA	PDRNN[25]	19.2
FNN	DBN and SVR[30]	21.2
NN	CNN[5]	7.7
LSTM	LSTM-CNN[31]	4.4
DNN	LSTM-CNN[31]	7.5
CNN	CNN with k-means[32]	12.3
MLP	Auto encoder and LSTM[33]	5.51

Table 13.4 Performance of the DL-based models and the ARIMA benchmark model.

	German data (2186 record)		Malaysian data (17,518 record)	
Methods	**RMSE**	**Runtime(s)**	**RMSE**	**Runtime(s)**
ARIMA	0.201	179.89	0.102	451.12
DNN	0.25	80.35	0.128	199.12
LSTM	0.197	431.11	0.097	902.56
LSTM-CNN	0.207	180.22	0.053	487.33

Some of these results are presented with a particular comparative context because the comparison of forecast results becomes more meaningful. In Reference 31, various algorithms and a new model of combined DL, which is a combination of long short-term memory (LSTM) and Convolutional Neural Network (CNN), is presented for STLF. Results show that DNN-based models are good candidates for use as STLF tools. The results for comparing the performance of DL models are presented in Table 13.4.

As shown in Table 13.4, the models' performance on different data structures is different. But in general, DL-based models have better performance than other models. Also, by increasing the input data, the performance of DL-based models is improved. The LSTM-CNN hybrid model has performed well on Malaysian data in terms of accuracy. The LSTM model also performed well on German load data. The DNN model performs very well in speed but has the worst accuracy performance among other DL models. In general, the LSTM-CNN hybrid model is faster than the LSTM model due to the use of CNN to extract features before forecasting with the LSTM model.

6. Conclusion

Electric load, particularly in household consumers due to dependency on various factors such as time and weather, is uncertain. Improving LF accuracy and speed has become an essential milestone in LF research. Proper selection of input parameters, models, and algorithms significantly impacts LF performance. Using DL-based methods in LF can be a proficient solution. This chapter introduced various aspects of LF, such as benefits, influencing factors, and challenges. Then LF classification and accuracy evaluation criteria of the models were introduced. Focusing on one of the most critical and practical components of LF, namely, forecasting models, the well-known DL models used in this field were introduced in the presence of existing uncertainties. Finally, several methods are compared in terms of the RMSE reduction percentage. Many researches have been done on multiple DL models to improve LF performance. Two or more DL models combined have been proposed to provide sufficient forecasting accuracy in some studies. Future research will focus on presenting and comparing hybrid models with several methods to address the shortcomings of current forecasting models.

References

1. Bunn DW. Forecasting loads and prices in competitive power markets. *Proc IEEE*. 2000; 88(2):163–169.
2. Hippert HS, Pedreira CE, Souza RC. Neural networks for short-term load forecasting: a review and evaluation. *IEEE Trans Power Syst*. 2001;16(1):44–55.
3. Tripathy DS, Prusty BR. Quantile regression averaging-based probabilistic forecasting of daily ambient temperature. *Int J Numer Model Electron Network Dev Field*. 2021; 34(3):e2846.
4. Tripathy DS, Prusty BR, Jena D. Short-term PV generation forecasting using quantile regression averaging. In: *2020 IEEE International Conference on Power Systems Technology (POWERCON)*. IEEE; September, 2020:1–6.
5. Goodfellow I, Bengio Y, Courville A. *Deep Learning*. MIT press; 2016.
6. Oprea SV, Bara A. Machine learning algorithms for short-term load forecast in residential buildings using smart meters, sensors and big data solutions. *IEEE Access*. 2019;7: 177874–177889.
7. Ruzic S, Vuckovic A, Nikolic N. Weather sensitive method for short term load forecasting in electric power utility of Serbia. *IEEE Trans Power Syst*. 2003;18(4): 1581–1586.
8. Chow JH, Wu FF, Momoh JA. Applied mathematics for restructured electric power systems. In: *Applied Mathematics for Restructured Electric Power Systems*. Boston, MA: Springer; 2005:1–9.
9. Al Mamun A, Sohel M, Mohammad N, Sunny MSH, Dipta DR, Hossain E. A comprehensive review of the load forecasting techniques using single and hybrid predictive models. *IEEE Access*. 2020;8:134911–134939.
10. Haida T, Muto S. Regression-based peak load forecasting using a transformation technique. *IEEE Trans Power Syst*. 1994;9(4):1788–1794.
11. Gao W, Darvishan A, Toghani M, Mohammadi M, Abedinia O, Ghadimi N. Different states of multi-block based forecast engine for price and load prediction. *Int J Electr Power Energy Syst*. 2019;104:423–435.
12. Sáez D, Ávila F, Olivares D, Cañizares C, Marín L. Fuzzy prediction interval models for forecasting renewable resources and loads in microgrids. *IEEE Trans Smart Grid*. 2014; 6(2):548–556.
13. Liu G, Tomsovic K. Quantifying spinning reserve in systems with significant wind power penetration. *IEEE Trans Power Syst*. 2012;27(4):2385–2393.
14. Manjili YS, Vega R, Jamshidi MM. Data-AnalyticBased adaptive solar energy forecasting framework. *IEEE Syst J*. March, 2018;12(1):285–296. https://doi.org/10.1109/JSYST.2017.2769483.
15. Li G, Chiang H-D. Toward cost-oriented forecasting of wind power generation. *IEEE Trans Smart Grid*. July, 2018;9(4):2508–2517. https://doi.org/10.1109/TSG.2016.2614341.

16. Katzir L, Loewenstern Y, Shmilovitz D. The effect of system characteristics on very-short-term load forecasting. *Przeglad Elektrotechniczny*. 2015;91(11):119–123.
17. Haq MR, Ni Z. A new hybrid model for short-term electricity load forecasting. *IEEE Access*. 2019;7:125413–125423.
18. Nazih AS, Fawwaz E, Osama M,A. Medium-term electric load forecasting using multi-variable linear and nonlinear regression. *Smart Grid Renew Energy*. 2011;2011.
19. Ye S, Zhu G, Xiao Z. Long term load forecasting and recommendations for China based on support vector regression. *Energy Power Eng*. 2012;4(5):380–385.
20. Hong T, Fan S. Probabilistic electric load forecasting: a tutorial review. *Int J Forecast*. 2016;32(3):914–938. https://doi.org/10.1016/j.ijforecast.2015.11.011.
21. Dechter R. Learning while searching in constraint-satisfaction-problems. In: *Proceedings of the Fifth AAAI National Conference on Artificial Intelligence, Philadelphia, Pennsylvania*; 1986:178–183. Accessed April 25, 2020.
22. Kuang P, Cao W, Wu Q. Preview on structures and algorithms of deep learning. In: *2014 11th International Computer Conference on Wavelet Active Media Technology and Information Processing(ICCWAMTIP)*. 2014:176–179. https://doi.org/10.1109/ICCWAMTIP.2014.7073385.
23. Ryu S, Noh J, Kim H. Deep neural network based demand side short term load forecasting. *Energies*. 2017;10(1):3. https://doi.org/10.3390/en10010003.
24. He Y, Deng J, Li H. Short-term power load forecasting with deep belief network and copula models. In: *2017 9th International Conference on Intelligent Human-Machine Systems and Cybernetics (IHMSC)*. Vol. 1. IEEE; August, 2017:191–194.
25. Shi H, Xu M, Li R. Deep learning for household load forecasting—a novel pooling deep RNN. *IEEE Trans Smart Grid*. 2017;9(5):5271–5280.
26. Kong W, Dong ZY, Hill DJ, Luo F, Xu Y. Short-term residential load forecasting based on resident behaviour learning. *IEEE Trans Power Syst*. 2017;33(1):1087–1088.
27. Sehovac L, Grolinger K. Deep learning for load forecasting: sequence to sequence recurrent neural networks with attention. *IEEE Access*. 2020;8:36411–36426.
28. Marino DL, Amarasinghe K, Manic M. Building energy load forecasting using deep neural networks. In: *IECON 2016-42nd Annual Conference of the IEEE Industrial Electronics Society*. IEEE; October, 2016:7046–7051.
29. Pai PF, Hong WC. Support vector machines with simulated annealing algorithms in electricity load forecasting. *Energy Convers Manag*. 2005;46(17):2669–2688.
30. Qiu X, Zhang L, Ren Y, Suganthan PN, Amaratunga G. Ensemble deep learning for regression and time series forecasting. In: *2014 IEEE Symposium on Computational Intelligence in Ensemble Learning (CIEL)*. IEEE; December, 2014:1–6.
31. Farsi B, Amayri M, Bouguila N, Eicker U. On short-term load forecasting using machine learning techniques and a novel parallel deep LSTM-CNN approach. *IEEE Access*. 2021;9:31191–31212.
32. Dong X, Qian L, Huang L. Short-term load forecasting in smart grid: a combined CNN and K-means clustering approach. In: *2017 IEEE International Conference on Big Data and Smart Computing (BigComp)*. IEEE; February, 2017:119–125.
33. Gensler A, Henze J, Sick B, Raabe N. Deep Learning for solar power forecasting—an approach using AutoEncoder and LSTM Neural Networks. In: *2016 IEEE International Conference on Systems, Man, and Cybernetics (SMC)*. IEEE; October, 2016:002858–002865.

Index

Note: Page numbers followed by "f" indicate figures and "t" indicate tables.

E

FSC
www.fsc.org
MIX
Papier | Fördert
gute Waldnutzung
FSC® C083411

Kolibri 360 GmbH
Ferdinand-Jühlke-Straße 7
99095 Erfurt, Deutschland
produktsicherheit@kolibri360.de